Got Your *New* IAP/SanTech *Catalog* Yet??

Fill out the card below and we'll get one to you fast!
While you're at it, give one to a friend!

IAP, Inc. is a major supplier to the Aviation Trade, featuring Quality Pilot supplies, including Plotters, Kneeboards, E-6Bs, Instrument Hoods, study guides, reference books, maintenance training materials and videos.

**Contact Your Nearest Dealer/ Distributor or call:
1-800-443-9250**

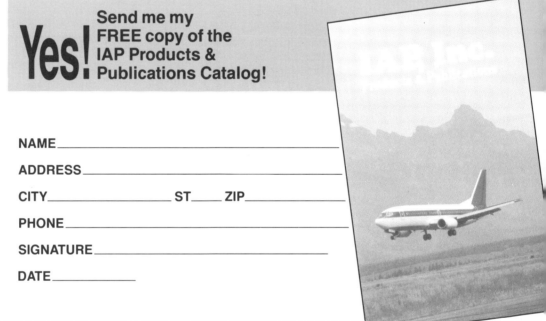

Yes! Send me my FREE copy of the IAP Products & Publications Catalog!

NAME_____
ADDRESS_____
CITY_____ ST___ ZIP_____
PHONE_____
SIGNATURE_____
DATE_____

Yes! Send me my FREE copy of the IAP Products & Publications Catalog!

NAME_____
ADDRESS_____
CITY_____ ST___ ZIP_____
PHONE_____
SIGNATURE_____
DATE_____

Call Toll Free
1-800-443-9250

For More Information On
Quality Pilot Products

BUSINESS REPLY MAIL
FIRST CLASS   PERMIT #4   CASPER, WY 82602-9990

POSTAGE WILL BE PAID BY IAP, INC.

IAP, Inc.
P.O. Box 10000
Casper, WY 82602-9927

BOOK ORDER
DEPARTMENT

NO POSTAGE
NECESSARY
IF MAILED
IN THE
UNITED STATES

BUSINESS REPLY MAIL
FIRST CLASS   PERMIT #4   CASPER, WY 82602-9990

POSTAGE WILL BE PAID BY IAP, INC.

IAP, Inc.
P.O. Box 10000
Casper, WY 82602-9927

BOOK ORDER
DEPARTMENT

NO POSTAGE
NECESSARY
IF MAILED
IN THE
UNITED STATES

ORDER NUMBER EA-AC 00-45C

AVIATION WEATHER SERVICES

Revised 1985

DEPARTMENT OF TRANSPORTATION
FEDERAL AVIATION ADMINISTRATION
Office of Flight Operations

DEPARTMENT OF COMMERCE
NATIONAL OCEANIC AND ATMOSPHERIC ADMINISTRATION
National Weather Service

Washington, D.C.

International Standard Book Number 0-89100-170-0
For sale by: IAP, Inc.
Mail to: P.O. Box 10000, Casper, WY 82602-1000
Ship to: 7383 6WN Road, Casper, WY 82604-1835
(800) 443-9250 • (307) 266-3838 • FAX: 307-472-5106

IAP, Inc.
7383 6WN Road, Casper, WY 82604-1835

© 1988 by IAP, Inc.
All Rights Reserved
Printed in the USA

PREFACE

AC 00-45C, Aviation Weather Services, is published jointly by the FAA Office of Flight Operations and the National Weather Service (NWS).

AC 00-45C, Aviation Weather Services, supplements the companion manual **AC 00-6A, Aviation Weather.** AC 00-6A deals with weather theories and hazards. The two manuals are sold separately; so at a nominal cost, a pilot can purchase a copy of this supplement (AC 00-45C) periodically and keep current in aviation weather services.

Clifton W. Green, National Weather Service Coordinator and Training Consultant at the FAA Academy, directed the preparation of AC 00-45C. John W. Victory and Phyllis Polland, meteorologists on the NWS Coordinator's staff, did the writing and editing of the manuscript. Recognition is given to Charles Lambert, who supplied text and illustrations for the International Services, and to Jerry Uecker, who supplied illustrations for domestic Aviation Weather Services. Recognition is given to John Blasic, NWS Liaison to FAA Headquarters, and the following meteorologists on the NWS Coordinator's staff for their comments and suggestions—David L. Carlson, Herschel T. Knowles, Richard A. Mitchem, Jon Osterberg, and James K. Purpura.

CONTENTS

	Page
ILLUSTRATIONS	vii
TABLES	x
INTRODUCTION	xiii

Section 1—THE AVIATION WEATHER SERVICE PROGRAM

Observations	1-1
National Oceanic and Atmospheric Administration	1-2
Service Outlets	1-8
Communication Systems	1-16
Users	1-16

Section 2—SURFACE AVIATION WEATHER REPORTS

Station Designator	2-1
Type and Time of Report	2-1
Sky Condition and Ceiling	2-1
Visibility	2-6
Weather and Obstructions to Vision	2-6
Sea Level Pressure	2-7
Temperature and Dew Point	2-7
Wind	2-8
Altimeter Setting	2-8
Remarks	2-8
Report Identifiers	2-15
Reading the Surface Aviation Weather Report	2-16
Automated Surface Observations	2-16

Section 3—PILOT AND RADAR REPORTS AND SATELLITE PICTURES

Pilot Weather Reports (PIREPS)	3-1
Radar Weather Reports (RAREPS)	3-2
Satellite Weather Pictures	3-6

Section 4—AVIATION WEATHER FORECASTS

Terminal Forecasts (FT and TAF)	4-1
Domestic Area Forecast (FA)	4-9
TWEB Route Forecasts and Synopsis	4-13
Inflight Advisories (WST, WS, WA)	4-14
Winds and Temperatures Aloft Forecast (FD)	4-18
Special Flight Forecast	4-18
Center Weather Service Unit Products (MIS and CWA)	4-19
Hurricane Advisory (WH)	4-20
Convective Outlook (AC)	4-20
Severe Weather Watch Bulletin (WW)	4-21

Section 5—SURFACE ANALYSIS CHART

Valid Time	5-1
Isobars	5-1
Pressure Systems	5-1
Fronts	5-1
Troughs and Ridges	5-3
Other Information	5-6
Using the Chart	5-6

Section 6—WEATHER DEPICTION CHART

Plotted Data	6-1
Analysis	6-1
Using the Chart	6-3

Section 7—RADAR SUMMARY CHART

Echo Type, Intensity, and Intensity Trend	7-1
Echo Configuration and Coverage	7-1
Echo Heights	7-4
Echo Movement	7-4
Severe Weather Watch Areas	7-4
Canadian Data	7-4
Using the Chart	7-5

Section 8—SIGNIFICANT WEATHER PROGNOSTICS

U.S. Low Level Significant Weather Prog	8-1
High Level Significant Weather Prog	8-5
International Flights	8-8

Section 9—WINDS AND TEMPERATURES ALOFT

Forecast Winds and Temperatures Aloft (FD)	9-1
Observed Winds Aloft	9-1
Using the Charts	9-5
International Flights	9-5

Section 10—COMPOSITE MOISTURE STABILITY CHART

Stability Panel	10-2
Freezing Level Panel	10-4
Precipitable Water Panel	10-4
Average Relative Humidity Panel	10-6
Using the Chart	10-7

Section 11—SEVERE WEATHER OUTLOOK CHART

General Thunderstorms	11-1
Severe Thunderstorms	11-1
Tornadoes	11-1
Using the Chart	11-1

Section 12—CONSTANT PRESSURE CHARTS

Plotted Data	12-1
Analysis	12-3
Three Dimensional Aspects	12-3
Using the Charts	12-10

Section 13—TROPOPAUSE DATA CHART

Observed Tropopause Panel	13-1
Using the Panel	13-1
Domestic Tropopause Wind and Wind Shear Progs	13-1
Using the Panels	13-6

Section 14—TABLES AND CONVERSION GRAPHS

Icing Intensities	14-1
Turbulence Intensities	14-1
Locations of Probable Turbulence by Intensities versus Weather and Terrain Features	14-1
Standard Conversions	14-3
Density Altitude Computation	14-4
Selected Contractions	14-5
Acronyms	14-7
Scheduled Issuance and Valid Times of Forecast Products	14-8

ILLUSTRATIONS

Figure   Page

Section 1—THE AVIATION WEATHER SERVICE PROGRAM

Figure	Title	Page
1-1.	Data flow in the aviation weather network	1-1
1-2.	The radar observing network	1-3
1-2A.	The Radar Remote Weather Display System network	1-4
1-3.	The forecast wind and temperatures aloft network	1-5
1-3A.	The forecast winds and temperatures aloft for Alaska and Hawaii	1-6
1-4.	Locations of WSFOs and airports for which each prepares terminal forecasts	1-7
1-4.	Continued	1-8
1-4A.	Locations of WSFOs in Alaska and Hawaii and airports for which each prepares terminal forecasts	1-9
1-5.	Locations of the area forecasts	1-10
1-5A.	Locations of the area forecasts in Alaska and Hawaii	1-11
1-6.	TWEB routes	1-12
1-6.	Continued	1-13
1-7.	Cross county TWEB routes	1-14
1-8.	Enroute Flight Advisory Service Facilities	1-15

Section 2—SURFACE AVIATION WEATHER REPORTS

Figure	Title	Page
2-1.	Scattered sky cover by a single advancing layer	2-2
2-2.	Scattered sky cover by a single layer surrounding the station	2-3
2-3.	Summation of cloud cover in multiple layers	2-4
2-4.	Summation of cloud cover in multiple layers	2-4
2-5.	The rotating beam ceilometer	2-5
2-6.	Vertical visibility	2-5
2-7.	Difference between runway visibility and runway visual range	2-9
2-8.	The transmissometer	2-9
2-9.	Towering Cumulus	2-12
2-10.	Cumulonimbus	2-12
2-11.	Cumulonimbus Mamma	2-13
2-12.	Altocumulus Castellanus	2-13
2-13.	Virga	2-14
2-14.	Standing Lenticular Altocumulus	2-14
2-15.	Decoding observations from unstaffed AMOS stations	2-17
2-16.	Decoding observations from staffed AMOS stations	2-17
2-17.	Decoding observations from AUTOB stations	2-19
2-18.	Decoding observations from RAMOS stations	2-20

Figure   Page

Section 3—PILOT AND RADAR REPORTS AND SATELLITE PICTURES

3-1.	Pilot Reports Format	3-1
3-2.	Digital Radar Report Plotted on a PPI Grid Overlay Chart	3-5
3-3.	Teletypewriter Plot of Echo Intensities for the South Central United States	3-6
3-4.	GOES Visible Imagery	3-7
3-5.	GOES Infra-red Imagery	3-8
3-6.	NOAA Visible Imagery	3-9
3-7.	NOAA Infra-red Imagery	3-10

Section 4—AVIATION WEATHER FORECASTS

4-1.	Area of jetstream turbulence	4-12
4-2.	Inflight Weather Advisory Location Identifiers (VORs)	4-15
4-3.	Geographical areas and terrain features	4-16

Section 5—SURFACE ANALYSIS

5-1.	Surface Weather Analysis Chart	5-2
5-2.	List of symbols on surface analyses	5-3
5-3.	Station model and explanation	5-5
5-4.	Sky cover symbols	5-6
5-5.	Present weather	5-7
5-6.	Barometer tendencies	5-8
5-7.	Cloud abbreviation	5-9

Section 6—WEATHER DEPICTION CHART

6-1.	A Weather Depiction Chart	6-2

Section 7—RADAR SUMMARY CHART

7-1.	A Radar Summary Chart	7-2

Section 8—SIGNIFICANT WEATHER PROGNOSTICS

8-1.	U.S. Low Level Significant Weather Prog	8-2
8-2.	U.S. Low Level 36 and 48 hour Significant Weather Prog	8-4
8-3.	U.S. High Level Significant Weather Prog	8-8
8-4.	International High Level Significant Weather Prog Chart	8-9

Section 9—WINDS AND TEMPERATURES ALOFT

9-1.	A panel of winds and temperatures aloft forecast for 24,000 feet pressure altitude	9-2
9-2.	An Observed Winds Aloft Chart	9-3
9-3.	A panel of observed winds aloft for 34,000 feet	9-4
9-4.	Polar stereographic forecast winds and temperatures aloft chart	9-6
9-5.	Mercator forecast winds and temperatures aloft chart	9-7

Figure   Page

Section 10—COMPOSITE MOISTURE STABILITY CHART

10-1.	Composite Moisture Stability Chart	10-1
10-2.	A Stability Panel of the Composite Moisture Stability Chart	10-2
10-3.	A Freezing Level Panel of the Composite Moisture Stability Chart	10-5
10-4.	A Precipitable Water Panel of the Composite Moisture Stability Chart	10-7
10-5.	Average Relative Humidity Panel of the Composite Moisture Chart	10-8

Section 11—SEVERE WEATHER OUTLOOK CHART

11-1.	Severe Weather Outlook Charts	11-2

Section 12—CONSTANT PRESSURE CHARTS

12-1.	Radiosonde Data Station Plot and Decode	12-1
12-2.	A section of an 850 millibar analysis	12-4
12-3.	A section of a 700 millibar analysis	12-5
12-4.	A section of a 500 millibar analysis	12-6
12-5.	A section of a 300 millibar analysis	12-7
12-6.	A section of a 200 millibar analysis	12-8
12-7.	A section of a 200 millibar analysis	12-9

Section 13—TROPOPAUSE DATA CHART

13-1.	Tropopause data chart	13-2
13-2.	An observed tropopause panel	13-3
13-3.	Section of a tropopause wind prog	13-4
13-4.	Section of a tropopause height/vertical wind shear prog	13-5

TABLES

Table   Page

Section 2—SURFACE AVIATION WEATHER REPORTS

2-1.	Summary of sky cover designators	2-2
2-2.	Ceiling designators	2-3
2-3.	Weather symbols and meanings	2-6
2-4.	Obstructions to vision—symbols and meanings	2-7
2-5.	Reportable visibility categories	2-18

Section 3—PILOT AND RADAR REPORTS AND SATELLITE PICTURES

3-1.	Precipitation intensity and intensity trend	3-3
3-2.	Ordered content of a radar weather report	3-3
3-3.	Contractions of radar operational status	3-4

Section 4—AVIATION WEATHER FORECASTS

4-1.	Categories	4-2
4-2.	Examples of categories groupings	4-2
4-3.	TAF weather codes	4-5
4-4.	Converting significant weather from U.S. terms to WMO terms	4-8
4-5.	Visibility conversion—TAF code to miles	4-9
4-6.	TAF cloud code	4-9
4-7.	TAF icing and turbulence	4-9
4-8.	Contractions in FAs	4-11
4-9.	Area coverage of showers and thunderstorms	4-11
4-10.	Variability terms	4-13

Section 5—SURFACE ANALYSIS

5-1.	Type of front	5-1
5-2.	Intensity of front	5-4
5-3.	Character of front	5-4

Section 6—WEATHER DEPICTION CHART

6-1.	Total sky cover	6-1
6-2.	Examples of plotting on the Weather Depiction Chart	6-3

Section 7—RADAR SUMMARY CHART

7-1.	Key to Radar Summary Chart	7-3

Table   Page

Section 8—SIGNIFICANT WEATHER PROGNOSTICS

8-1.	Some standard weather symbols	8-1
8-2.	Significant weather prognostic symbols	8-3
8-3.	Depiction of clouds and turbulence on a High Level Significant Weather Prog	8-7

Section 10—COMPOSITE MOISTURE STABILITY CHART

10-1.	Thunderstorm Potential	10-3
10-2.	Plotting freezing levels	10-4
10-3.	Vertical temperature profile of plotted freezing levels at a station	10-6

Section 11—SEVERE WEATHER OUTLOOK CHART

11-1.	Notation of Coverage	11-1

Section 12—CONSTANT PRESSURE CHARTS

12-1.	Features of constant pressure charts-U.S.	12-2
12-2.	Examples of radiosonde plotted data	12-2

Section 14—TABLES AND CONVERSION GRAPHS

14-1.	Icing intensities, airframe ice accumulation and pilot report	14-1
14-2.	Turbulence reporting criteria	14-2
14-3.	Scheduled issuance and valid times of forecast products	14-8

INTRODUCTION

The rapid expansion of air transportation makes necessary a move toward mass briefings to meet aviation demands. As a results, you, the pilot, must become increasingly self-reliant in getting your weather information. On occasion, you may need to rely entirely on self-briefing.

This advisory circular, AC 00-45C, explains weather service in general and the details of interpreting and using coded weather reports, forecasts, and weather charts, both observed and prognostic. Many charts and tables apply directly to flight planning and in-flight decisions.

This advisory circular is an excellent source of study for pilot certification examinations. Its 14 sections contain information needed by all pilots, from the student pilot to the airline transport pilot.

AC 00-45 is updated periodically to reflect changes brought about by the latest service demands, techniques, and capabilities. The purchase of an updated copy is a wise investment for any active pilot.

Comments and suggestions for improving this publication are encouraged and should be directed to:

National Weather Service Coordinator, AAC-909
Federal Aviation Administration
Mike Monroney Aeronautical Center
P.O. Box 25082
Oklahoma City, OK 73125

Advisory Circular, AC 00-45C, supersedes AC 00-45B, Aviation Weather Service, revised 1979.

Section 1
THE AVIATION WEATHER SERVICE PROGRAM

Weather service to aviation is a joint effort of the National Weather Service (NWS), the Federal Aviation Administration (FAA), the Department of Defense (DOD) weather service, and other aviation oriented groups and individuals. Because of international flights and a need of world-wide weather, foreign weather services also have a vital input into our service.

This section follows the development and flow of observations, reports, and forecasts through the service to the users, as depicted in figure 1-1.

OBSERVATIONS

Weather observations are measurements and estimates of existing weather both at the surface and aloft. When recorded and transmitted, an observation becomes a report and these reports are the basis of all weather analyses and forecasts. Note in figure 1-1 that high speed communications and automated data processing have improved the flow of weather reports to the aviation user.

Surface Observations

Surface aviation observations include weather elements pertinent to flying. A network of airport stations provides routine up-to-date aviation weather reports. Most of the stations in the network are either NWS or FAA; however, the military services and contracted civilians are also included. A major change in the surface weather observation network is underway with the installation of

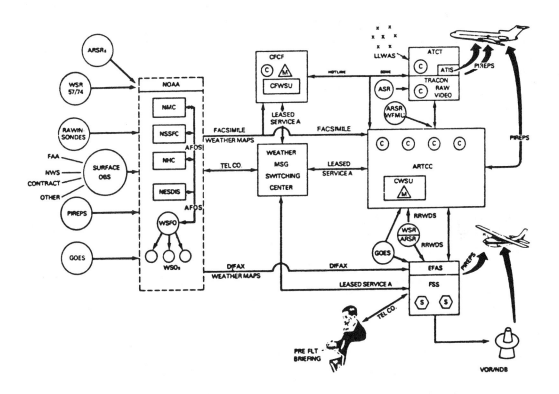

FIGURE 1-1. Data flow in the aviation weather network. Note the important feedback of pilot reports.

1-1

automated weather observing stations across the country. These automated stations are expected to become a major part of the network. Existing types of automated observations are discussed in Section 2.

Radar Observations

Precipitation reflects radar signals and the reflected signals are displayed as echoes on the radar scope. NWS radar covers nearly all the U.S. east of the Rocky Mountains. Radar coverage over the remainder of the U.S. is largely by Air Route Traffic Control radars. Thus, except for some western mountainous terrain, radar coverage is nearly complete over the contiguous 48 states. Figure 1-2 maps the radar observing network.

The new Radar Remote Weather Display System (RRWDS) is a significant improvement over previous radar remote systems. This system is specifically designed to provide *real-time* radar weather information from many different radars, which makes it very useful to the FSS specialist and the ARTCC meteorologist. The RRWDS display is similar to the color video display systems of private enterprise. It is connected to FAA and Air Force Air Route Surveillance radars as well as NWS weather radars (figure 1-2A). This gives briefers access to real-time radar weather information from areas of the country where it was previously not available.

Satellite Observations

Visible and infrared images of clouds are available from weather satellites in orbit. Satellite pictures are an important additional source of weather observations. GOES and NOAA satellite products are available through the facsimile network and directly from NWS Satellite Field Service Stations (SFSS). For more information on satellite products, see Section 3 "Satellite Pictures".

Upper Air Observations

Another important source of observed weather data is from radiosonde balloons and PIREPs. Upper air observations from radiosonde taken twice daily at specified stations furnish temperature, humidity, pressure, and wind, often to heights above 100,000 feet. Pilots themselves are a vital source of upper air weather observations. In fact, aircraft in flight are the only means of directly observing turbulence, icing, and the height of cloud tops.

Low Level Wind Shear Alert System (LLWAS)

This system provides pilots and controllers with information on hazardous surface wind conditions (on or near the airport) that creates unsafe landing or departure conditions. The system is a real-time, computer controlled, surface wind sensor system which evaluates wind speed and direction from sensors on the airport periphery with center field wind data. During the time that an alert is posted, air traffic controllers provide wind shear advisories to all arriving and departing aircraft.

NATIONAL OCEANIC AND ATMOSPHERIC ADMINISTRATION (NOAA)

NOAA collects and analyzes data and prepares forecasts on a national, hemispheric, and global basis. Following is a description of those facilities tasked with this duty.

National Environmental Satellite Data and Information Service (NESDIS)

The National Environmental Satellite Data and Information Service (NESDIS) located in Washington D.C. directs the weather satellite program and works in close cooperation with NWS meteorlogists at National Meteorological Center (NMC) and the Satellite Field Service Stations (SFSS). Satellite cloud photographs are available at field stations from NMC via facsimile or directly from a SFSS. Figures 3-4 and 3-5 are examples of GOES satellite pictures received from a SFSS.

National Meteorological Center (NMC)

The National Meteorological Center (NMC) of the NWS, located in Washington D.C. is the hub of weather processing. From worldwide weather reports it prepares guidance forecasts and charts of observed and forecast weather for use by various forecast facilities as described below. Many of the charts are computer prepared. Others are computer outputs adjusted and annotated by meteorologists. A few are manually prepared by forecasters.

Some NMC products are specifically for aviation. For example, NMC prepares the wind and temperatures aloft forecast. Figure 1-3 is the network of forecast winds and temperatures for the contiguous 48 states. Figure 1.3A shows the Alaskan and Hawaiian network of forecast winds and temperatures.

National Hurricane Center (NHC)

The NWS National Hurricane Center (NHC) located in Miami FL develops hurricane forecasting techniques and issues hurricane forecasts for the Atlantic, the Caribbean, the Gulf of Mexico and adjacent land areas. Hurricane warning centers at San Francisco and Honolulu issue warnings for the eastern and central Pacific.

National Severe Storms Forecast Center (NSSFC)

The NWS National Severe Storms Forecast Center (NSSFC) issues forecasts of severe convective

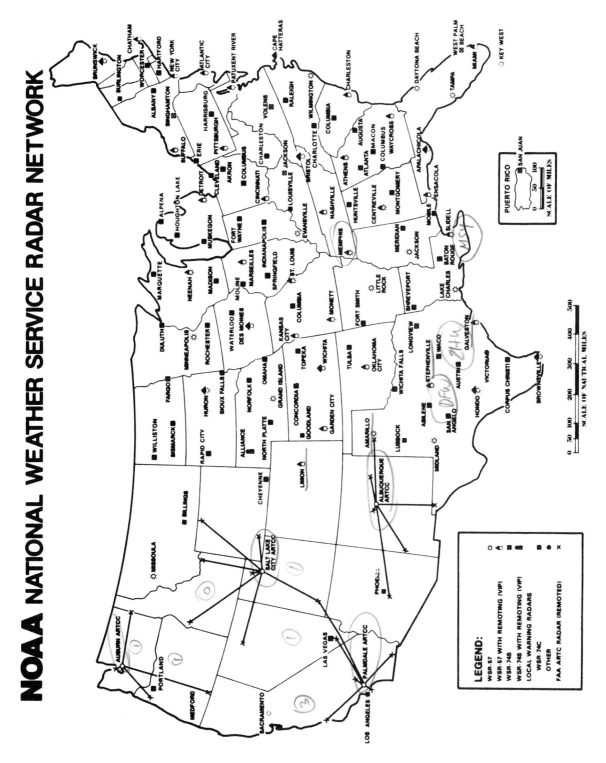

FIGURE 1-2. The radar observing network.

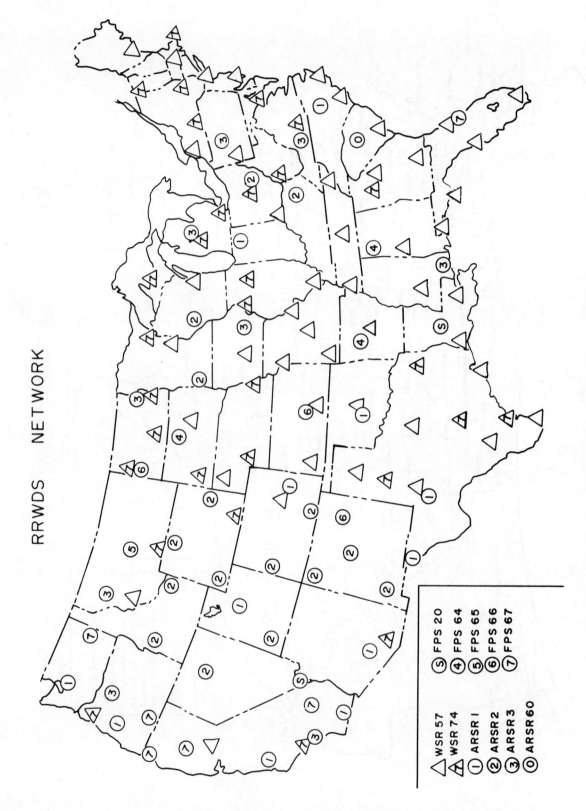

FIGURE 1-2A. The Radar Remote Weather Display System (RRWDS) network.

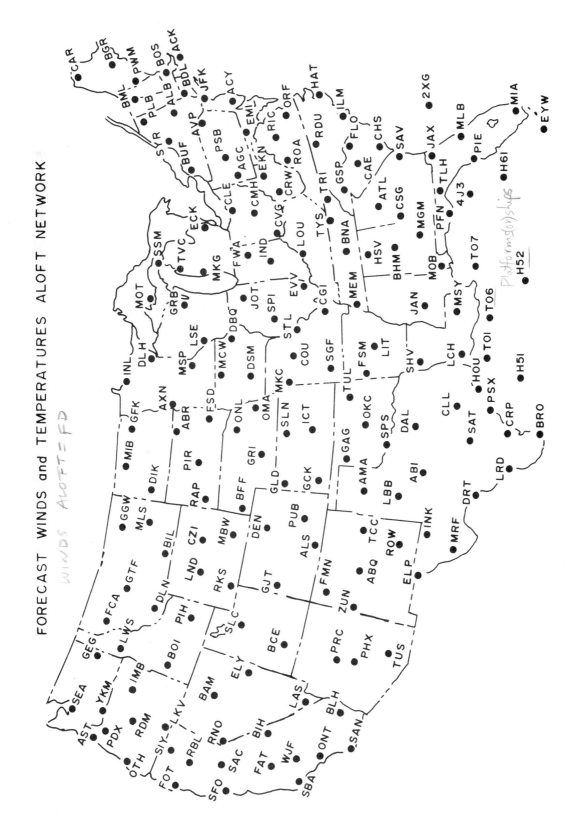

FIGURE 1-3. The forecast winds and temperatures aloft network.

FD LOCATIONS - ALASKA and HAWAII

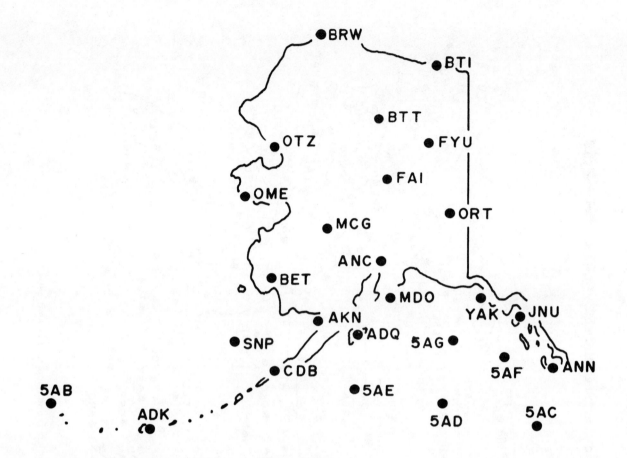

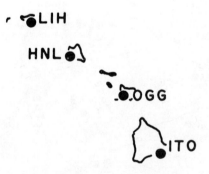

FIGURE 1-3A. The forecast winds and temperatures aloft network for Alaska and Hawaii.

storms, such as severe weather watches and convective outlooks, for the contiguous 48 states. It is located at Kansas City MO near the heart of the area most frequently affected by severe thunderstorms.

National Aviation Weather Advisory Unit (NAWAU)

This is a National Weather Service aviation dedicated unit located in Kansas City MO. Meteorologists in this unit prepare and issue area forecasts and inflight advisories (convective and non-convective SIGMETS and AIRMETS) for the contiguous 48 states (figure 1-5). These two products were formerly prepared and issued by designated WSFOs.

Weather Service Forecast Office (WSFO)

A Weather Service Forecast Office (WSFO) issues various public and aviation oriented forecasts and weather warnings for their area of responsibility. In support of aviation, products include terminal forecasts as well as Transcribed Weather Broadcast (TWEB) synopses and route forecasts. Figure 1-4 and 1-4A show locations of WSFOs and the airports for which each office prepares terminal forecasts. Figures 1-6 and 1-7 show TWEB routes.

Weather Service Office (WSO)

A Weather Service Office (WSO) prepares and issues public forecasts and warnings and provides general weather service for their local areas. A WSO

FIGURE 1-4. Locations of WSFOs, their areas of responsibility, and airports for which each prepares terminal forecasts.

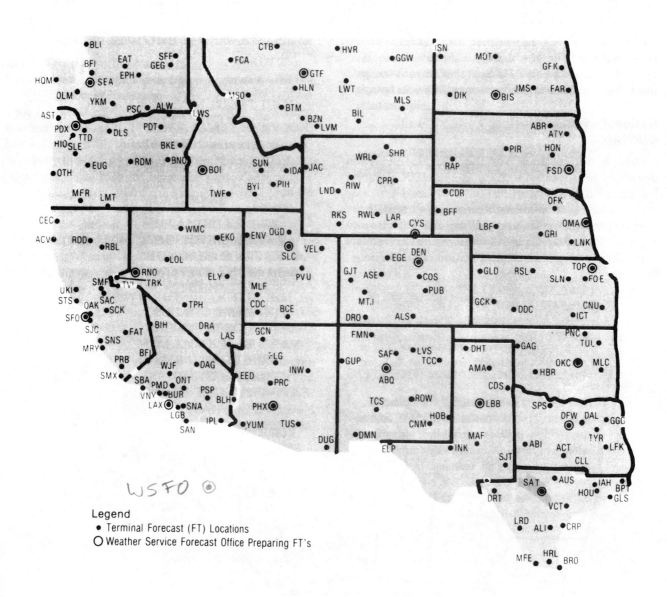

FIGURE 1-4. Continued.

may also amend terminal forecasts for a period of two hours or less when unexpected significant changes in the weather occur.

SERVICE OUTLETS

A weather service outlet is defined as any facility, either government or non-government, that provides aviation weather service. This section discusses only FAA and NWS outlets.

Flight Service Stations (FSS)

The FAA Flight Service Station (FSS) provides more aviation weather briefing service than any other government service outlet. It provides preflight and inflight briefings, makes scheduled and unscheduled weather broadcasts, and furnishes weather advisories to known flights in the FSS area. Because of the tremendous number of flight operations, selected FSSs also provide transcribed weather briefings. By listening to the recordings, you can assess further need for more detailed person to person briefing. There are two types of recordings - (1) Transcribed Weather Broadcast (TWEB) and (2) Pilot's Automatic Telephone Weather Answering Service (PATWAS).

The TWEB is a continous broadcast on low/medium frequencies (200 to 415 kHz) and selected VORs (108.0 to 117.95 MHz). The TWEB is based on a route of flight concept with the order and content of the TWEB transcription as follows:

1. Synopsis
2. Flight Precautions
3. Route Forecasts
4. Outlook (Optional)

1-8

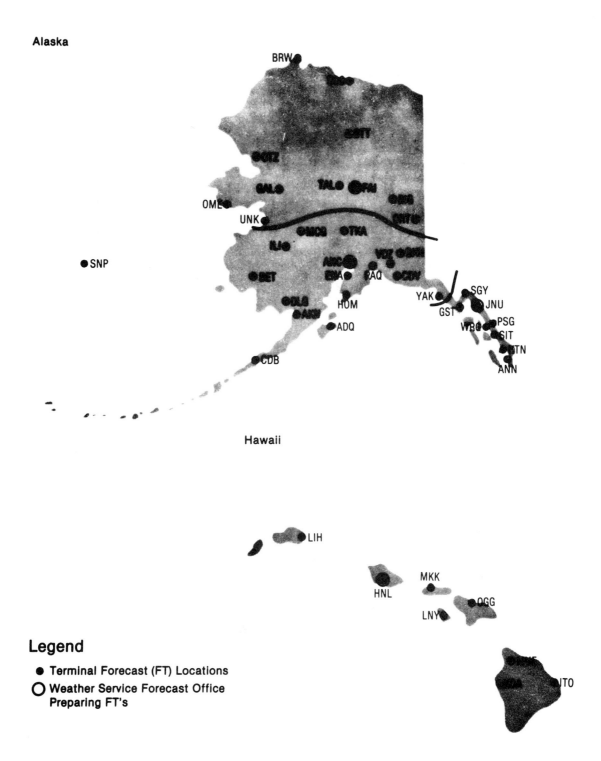

FIGURE 1-4A. Locations of WSFOs in Alaska and Hawaii, their areas of responsibility, and airpots for which each prepares terminal forecasts.

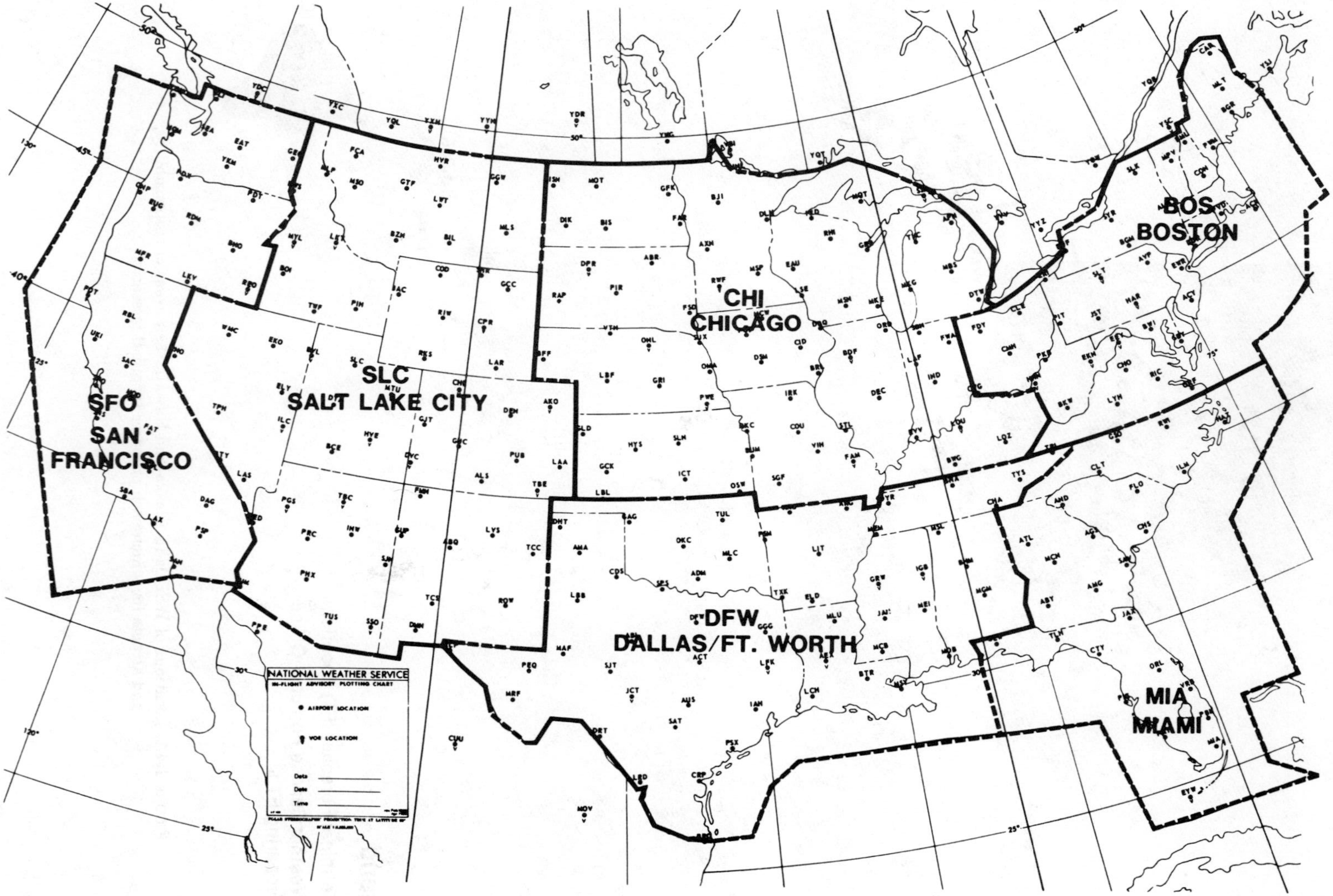

FIGURE 1-5. Locations of the area forecasts.

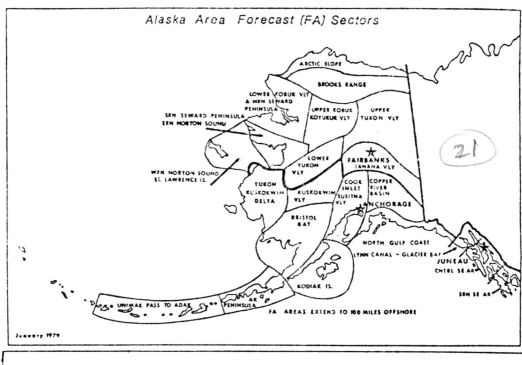

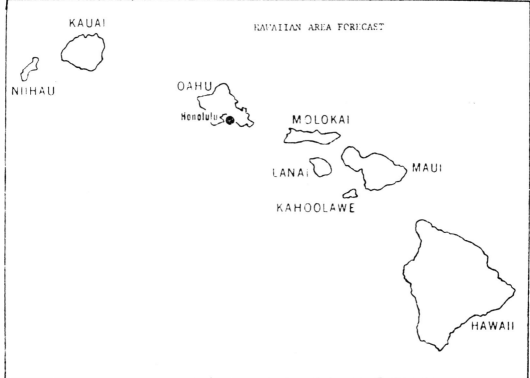

Figure 1-5A. Locations of the area forecasts in Alaska and Hawaii.

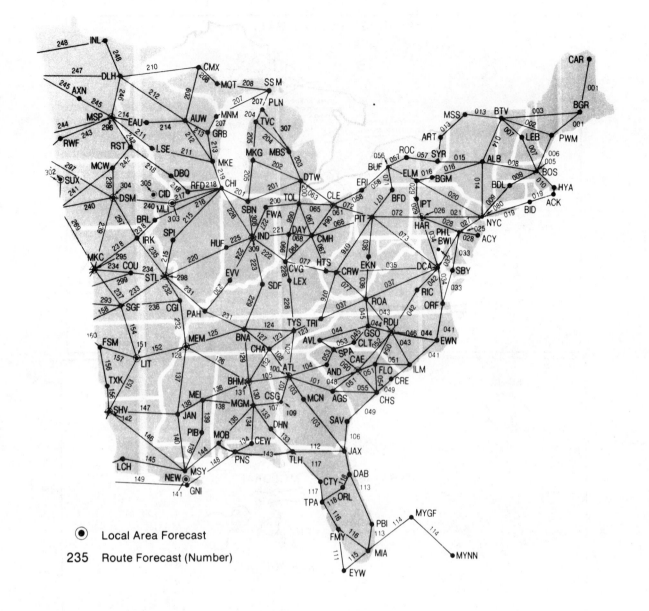

FIGURE 1-6. Numbered routes for which TWEB route forecasts are prepared.

5. Winds Aloft Forecast
6. Radar Reports
7. Surface Weather Reports
8. Pilot Reports
9. Notice to Airmen (NOTAMs)

The first five items are forecasts prepared by the NWS and are discussed in detail in Section 4. The synopsis and route forecast are prepared specifically for the TWEB by the WSFOs. Flight precautions, outlook and winds aloft are adapted respectively from inflight advisories, area forecasts and the NMC winds aloft forecast. Radar reports and pilot reports are discussed in Section 3. Surface reports are the subject of Section 2.

PATWAS is a recorded telephone briefing service with the forecast for the local area - usually within a 50 nautical mile radius of the station. A few selected stations also include route forecasts similar to the TWEB.

The order and content of the PATWAS recording are as follows:

1. Introduction (describing PATWAS area)
2. Adverse Conditions
3. Recommendation (VFR flight not recommended, if appropriate)
4. Synopsis
5. Current Conditions
6. Surface Winds

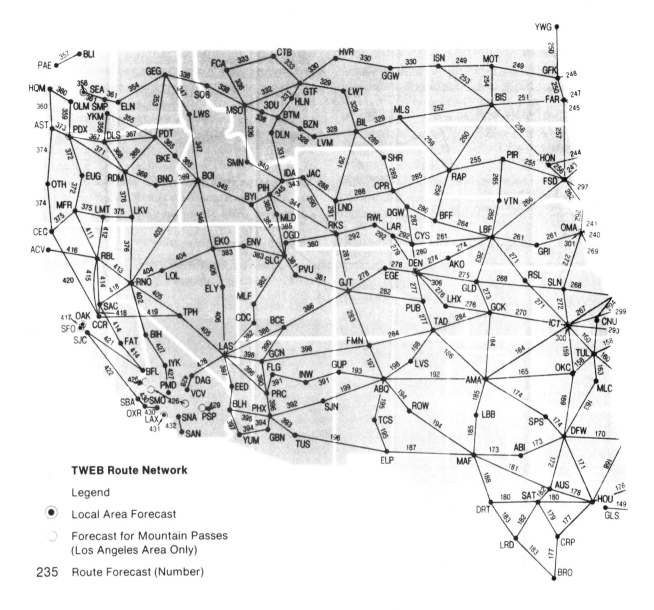

FIGURE 1-6. Continued.

7. Forecast
8. Winds Aloft
9. NOTAMS
10. Military Training Activity
11. Closing Announcements

FAA facilities providing PATWAS have operational procedures that place a high operational priority on PATWAS. This insures the information is current and accurate. Detailed PATWAS information is usually prepared at selected time intervals between 0500 and 2200 local time with updates issued as needed. A general outlook for the PATWAS area is available between 2200 and 0500 local time.

Figure 1-6 shows TWEB routes for which forecasts are prepared. Figure 1-7 shows cross-country TWEB routes. The Airman's Information Manual gives basic flight information and Air Traffic Control (ATC) procedures. The Airport Facility Directory lists PATWAS telephone numbers of FSS and NWS briefing offices.

The Hazardous In-flight Weather Advisory Service (HIWAS) is a continuous broadcast service of in-flight weather urgent PIREPs, CWAs, and AWWs over selected VORs. Also, hazardous weather not yet covered by an advisory will be included. In areas where HIWAS is already being utilized, controllers and specialists have discontinued their routine broadcast of in-flight advisories but continue broadcasting a short alerting message.

The Enroute Flight Advisory Service (Flight Watch) is a weather service on a common frequency

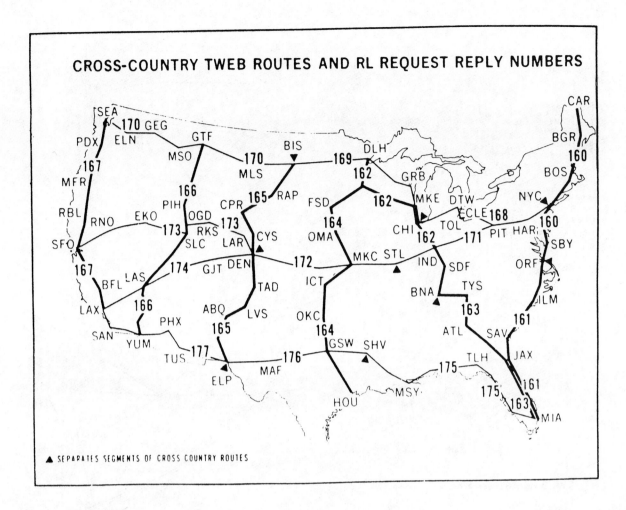

FIGURE 1-7. Cross country numbered routes for which route forecasts are available.

of 122.0 MHz from selected FSSs. The Flight Watch specialist maintains a continuous weather watch, provides time-critical assistance to enroute pilots facing hazardous or unknown weather, and may recommend alternate or diversionary routes. Additionally, Flight Watch is a focal point for rapid receipt and dissemination of pilot reports. Figure 1-8 indicates the sites where EFAS and associated outlets are located. To avail yourself of this service, call "FLIGHT WATCH" on 122.0 MHz (example "JACKSONVILLE FLIGHT WATCH, THIS IS...".

Air Traffic Control Command Center (ATCCC)

This operational facility is located at FAA headquarters in Washington D.C. Its objective is to manage the flow of air traffic on a system-wide basis, to minimize delays by watching capacity and demand, and to achieve maximum utilization of the airspace. Because weather is the overwhelming reason for air traffic delays and reroutings, this facility is supported by full-time NWS meteorologists whose function is to advise ATCCC flow controllers by continuously monitoring the weather throughout the system and anticipating weather developments that might affect system operations.

Air Route Traffic Control Center (ARTCC)

An ARTCC is a radar facility established to provide air traffic control service to aircraft operating on IFR flight plans within controlled airspace and principally during the enroute phase of flight.

Center Weather Service Unit (CWSU)

All FAA facilities within an ARTCC boundary are supported by a CWSU. This unit is a joint agency aviation weather support team located at each ARTCC. The unit is composed of National Weather Service (NWS) meteorologists and FAA controllers, the latter being assigned as Weather Coordinators. The primary task of the CWSU meteorologist is to provide FAA facilities within the ARTCC area of responsibility with accurate and timely weather information. This information is based on a continuous analysis and interpretation of - (1) real-time weather data at the ARTCC through the use of radar, satellite and PIREPs and (2) various NWS products such

1-14

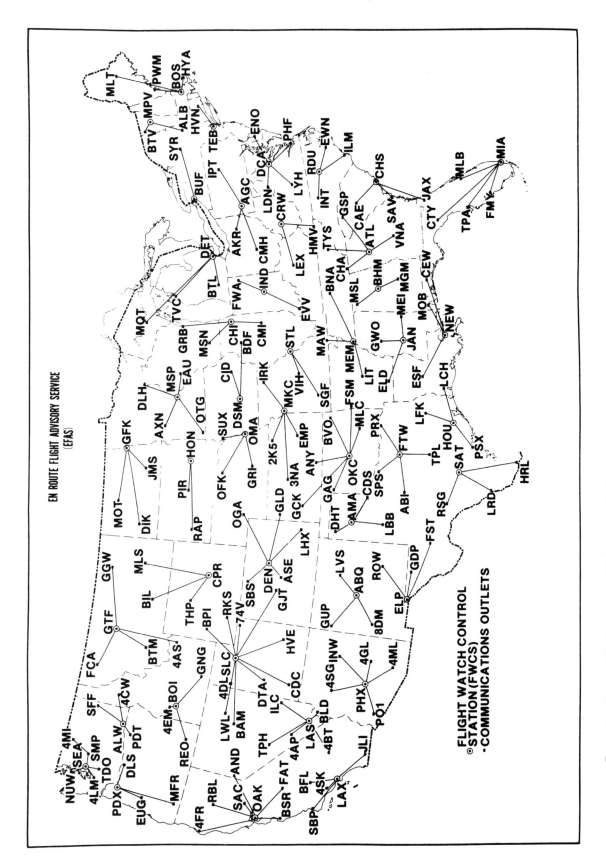

FIGURE 1-8. Enroute Flight Advisory (Flight Watch Facilities). An aircraft at 5,000 feet can receive a transmission to a distance of about 80 miles from any central or remote site.

as terminal and area forecasts, in-flight advisories, etc. The flow or exchange of weather information between the CWSU meteorologists and the FAA facilities is the primary task of the Weather Coordinator.

Similar to CWSU in the ARTCCs, there is a Central Flow Weather Service Unit (CFWSU) located in the Central Flow Control Facility (CFCF) in the ATCCC. The on-duty meteorologist in the CFWSU has the responsibility of the weather coordination on the national level.

Terminal Control Facility

The FAA terminal controller informs arriving and departing aircraft of pertinent local weather conditions. The controller becomes familiar with and remains aware of current weather information needed to perform air traffic control duties in the vicinity of the terminal. The responsibility for reporting visibility observations is shared with the NWS at many ATCT facilities. At other tower facilities, the controller has the full responsibility for observing, reporting, and classifying aviation weather elements.

Automatic Terminal Information Service (ATIS) is provided at most major airports to inform pilots, as they approach the terminal area, of the current weather and other pertinent local airport information.

Weather Service Office

NWS Offices provide weather briefings in areas not served by Flight Service Stations and provide local warnings to aviation. They furnish backup assistance to FAA service outlets.

Weather Service Forecast Office

NWS Forecast Offices provide some selective pilot briefings and supply backup service to FAA outlets. When getting a briefing from a FSS, you may, if necessary, request a telephone "patch in" to the WSFO forecaster. A few WSFOs also make and record PATWAS.

AM WEATHER

A fifteen minute weather program is broadcast Monday through Friday mornings nationally on approximately 250 Public Broadcast Television Stations.

Professional meteorologists from the National Weather Service and the National Environment Satellite Data and Information Service provide weather information primarily for pilots to enable them to make a better go or no-go flight decision.

National and Regional Weather Maps are provided, along with satellite sequences, radar reports, winds aloft, radar reports, and weatherwatches. Extended forecasts are provided daily and on Fridays to cover the weekend. AM WEATHER broadcasts also serve many other interest areas that depend upon forecasts.

The program draws upon the U.S. weather observation network, on geostationary and polar orbiting satellite data, and on computer analysis to produce daily forecasts with 85 to 90% accuracy.

COMMUNICATION SYSTEMS

As noted earlier, high speed communications and automated data processing have improved the flow of weather data and products through the aviation weather network. Examining figure 1-1, the flow of weather information between the NOAA weather facilities is accomplished through the Automation of Field Operations and Services (AFOS) communications system. Alphanumeric and graphic products are displayed on a cathode ray tube (CRT), similar to a television screen, which eliminates the need for the slower teletypewriters and facsimile machines.

The flow of alphanumeric weather information to the FAA Service Outlets is accomplished through the Leased Terminal Equipment (LTE) which also displays data on a CRT and so eliminates the need for teletypewriters. The Leased Terminal Equipment is replacing the Leased Service A System (LSAS) in Figure 1.1.

Exchange of weather information between the NWS and FAA Service Outlets is generally accomplished in two ways. Graphic products (weather maps) are received by FAA Service Outlets from NMC in Washington, D.C. over a low speed facsimile system. Alphanumeric information is exchanged through the Weather Message Switching Center (WMSC) in Kansas City, MO. This switching facility serves as the gateway for the flow of alphanumeric information from one communication system to another (i.e. between the various FAA facilities, NWS, and other users).

USERS

The ultimate users of the aviation weather service are pilots and dispatchers. Maintenance personnel also may use the service in protecting idle aircraft against storm damage. As a user of the service, you also contribute to it. Send pilot weather reports (PIREPs) to help your fellow pilots, briefers and forecasters. The service can be no better or more complete than the information that goes into it.

In the interest of safety, you should get a complete briefing before each flight. If you have L/MF radio, you can get a preliminary briefing by listening to the TWEB at your home or place of business. If you don't have a radio and PATWAS is available, dial PATWAS for a briefing. Many times the weather situation may be complex and you may not completely comprehend the recorded message. If you need additional information after listening to the TWEB or PATWAS, you should contact the FSS or WSO for a more complete briefing tailored for your specific flight.

How to Get a Good Weather Briefing

When requesting a briefing, make known you are a pilot. Give clear and concise facts about your flight:

1. Type of flight VFR or IFR
2. ACFT Ident or pilot's name
3. ACFT type
4. Departure point
5. Route-of-flight
6. Destination
7. Altitude
8. Estimated time of departure
9. Estimated time en route or EST time of arrival

With this background, the briefer can proceed directly with the briefing and concentrate on weather relevant to your flight.

The weather briefing you receive depends on the type requested. A STANDARD briefing should include:

1. Hazardous weather if any (you may elect to cancel at this point)
2. Weather synopsis (positions and movements of lows, highs, fronts, and other significant causes of weather)
3. Current weather
4. Forecast weather (enroute and destination)
5. Forecast winds aloft
6. Alternate routes (if any)
7. Aeronautical information (NOTAMs)

An ABBREVIATED briefing will be provided when the user requests information 1) to supplement mass disseminated data, 2) to update a previous briefing or 3) be limited to specific information.

An OUTLOOK briefing will be provided when the briefing is six or more hours in advance of proposed departure. Briefing will be limited to applicable forecast data for the proposed flight.

The FSSs and WSOs are to serve you. You should not hesitate to discuss factors that need elaboration or to ask questions. You have a complete briefing only when *you* have a clear picture of the weather to expect. It is to your advantage to make a final weather check immediately before departure if at all possible.

Request/Reply Service

The request/reply service is available at all FSSs, WSOs and WSFOs. You may request through the service any reports or forecasts not routinely available at your service outlet. These include route forecasts used in TWEB and PATWAS, recorder briefings and RADAR plots (see figure 3-3). You can request a forecast for any numbered route shown in figure 1-6 or any of the longer cross-country routes shown in figure 1.7.

Have an Alternate Plan of Action

When weather is questionable, get a picture of expected weather over a broader area. Preplan a route to take you rapidly away from the weather if it goes sour. When you fly into weather through which you cannot safely continue, you must act quickly. Without preplanning, you may not know the best direction to turn; a wrong turn could lead to disaster. A preplanned diversion beats panic. Better be safe than sorry.

Section 2
SURFACE AVIATION WEATHER REPORTS

When an observation is reported and transmitted, it is a weather *report*. A surface aviation weather report contains some or all of the following elements:

1. Station designator
2. Type and time of report
3. Sky condition and ceiling
4. Visibility
5. Weather and obstructions to vision
6. Sea level pressure
7. Temperature and dew point
8. Wind direction, speed, and character
9. Altimeter setting
10. Remarks and coded data

Those elements not occurring at observation time or not pertinent to the observation are omitted from the report. When an element should be included but is unavailable, the letter "M" is transmitted in lieu of the missing element. Those elements that are included are transmitted in the above sequence.

Following are five (5) reports as transmitted. These reports are used in discussing the above 10 elements. If you have this reference in a loose leaf binder, you will find it helpful to remove this page and keep it before you as you proceed through the discussion.

INK SA 1854 CLR 15 106/77/63/1112G18/000
BOI SA 1854 150 SCT 30 181/62/42/1304/015
LAX SA 1852 7 SCT 250 SCT 6HK 129/60/59/
   2504/991
MDW RS 1856 -X M7 OVC 1 1/2R+F
990/63/61/3205/
   980/RF2 RB12
JFK RS 1853 W5 X 1/4F 180/68/64/1804/006/
   R04RVR22V30 SFC VSBY 1/2

STATION DESIGNATOR

The station designator is the three-letter location identifier for the reporting station. These five reports are from Wink TX (INK), Boise ID (BOI), Los Angeles CA (LAX), Chicago Midway Airport IL (MDW), and John F. Kennedy Airport, New York City NY (JFK).

TYPE AND TIME OF REPORT

The two basic types of reports are:

1. Record observation (SA), reports taken on the hour and
2. Special reports (RS or SP), observations taken when needed to report significant changes in weather.

Record observations (SA) are transmitted in sequenced collectives and are indentified by sequence headings. The first three reports are of this type (INK, BOI and LAX). A record special is a record observation that reports a significant change in weather. It is identified by the letters "RS" as shown in the reports from MDW and JFK. A special "SP" is an observation taken other than on the hour to report a significant change in weather. All reports transmitted must convey the time in Greenwich Mean Time and the type of observation.

SKY CONDITION AND CEILING

A clear sky or a layer of clouds or obscuring phenomena *aloft* is reported by one of the first seven *sky cover designators* in table 2-1. A layer is defined as clouds or obscuring phenomena with the base at approximately the same level. Height of the base of a layer precedes the sky cover designator. Height is in hundreds of feet *above ground level*.

Note that INK is reporting sky clear. No height precedes the designator since no sky cover is reported. BOI reports a scattered layer at 15,000 feet above the station. Figures 2-1 and 2-2 illustrate single layers of scattered clouds.

When more than one layer is reported, layers are in ascending order of height. For each layer above a lower layer or layers, the sky cover designator for that layer represents the *total sky* covered by that layer and all lower layers. In other words, the summation concept of cloud layers is used. LAX reports two layers, a scattered layer at 700 feet and a higher layer at 25,000 feet. Total coverage of the two layers does not exceed 5/10 coverage, so the upper layer also is reported as scattered. Figure 2-3 and 2-4 illustrate cloud over of multiple layers.

TABLE 2-1. Summary of sky cover designators

Designator	Meaning	Spoken
CLR	Clear. (Less than 0.1 sky cover.)	CLEAR
SCT	Scattered layer Aloft. (0.1 through 0.5 sky cover.)	SCATTERED
BKN*	Broken Layer Aloft. (0.6 through 0.9 sky cover.)	BROKEN
OVC*	Overcast Layer Aloft. (More than 0.9, or 1.0 sky cover.)	OVERCAST
−SCT	Thin scattered. At least ½ of the sky cover aloft is transparent at and below the level of the layer aloft.	THIN SCATTERED
−BKN	Thin Broken.	THIN BROKEN
−OVC	Thin Overcast.	THIN OVERCAST
X* 1.0	Surface Based Obstruction. (All of sky is hidden by surface based phenomena.)	SKY OBSCURED
−X	Surface Based Partial Obscuration. (0.1 or more, but not all, of sky is hidden by surface based phenomena.	SKY PARTIALLY OBSCURED

(handwritten note: NOT A CEILING)

*Sky condition represented by this designator will constitute a ceiling layer.

"Transparent" sky cover is clouds or obscuring phenomena aloft through which blue sky or higher sky cover is visible. As explained in table 2-1, a scattered, broken or overcast layer may be reported as "thin". To be classified as thin, a layer must be half or more transparent, and remember that sky cover of a layer includes all sky cover below the layer. For example, if at LAX the sky had been visible through half or more of the total sky cover reported by the higher layer, the report would have been

LAX SA 1854 7 SCT 250-SCT etc.

Any phenomenon *based at the surface* and hiding all or part of the sky is reported as SKY OBSCURED* or SKY PARTIALLY OBSCURED* as explained in table 2-1. An obsuration or partial obscuration may be caused by precipitation, fog, dust, blowing snow, etc. No height value precedes the designator for partial obscurations since vertical visibility is not restricted overhead. A height value precedes the designator for a total obscuration and denotes vertical visibility into the phenomenon.

Ceiling is defined as:

1. Height of the lowest layer of clouds or obscuring phenomena aloft that is reported as broken or overcast and not classified as thin, or

*Descriptions in capital letters are the usual phraseology in which these reports are broadcast.

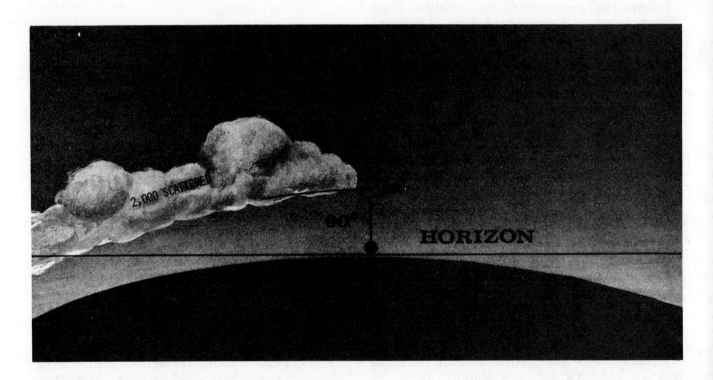

FIGURE 2-1. Scattered sky cover by a single advancing layer. Scattered is 5/10 or less sky cover (5/10 in this example).

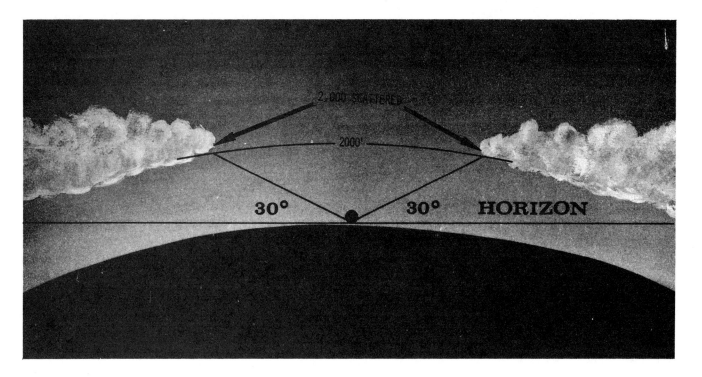

FIGURE 2-2. Scattered sky cover by a single layer surrounding the station (5/10 covered in this example).

2. Vertical visibility into surface-based obscuring phenomena that hides all the sky.

Now look at the reports from MDW and JFK. MDW reports a partial obscuration and an overcast at 700 feet. The overcast constitutes a ceiling at 700 feet. Note also that the height of this ceiling layer is preceded by the letter "M". JFK reports a total obscuration and the height value preceding the sky cover designator represents 500 feet vertical visibility into the obscuring phenomenon. Height of the ceiling value is preceded by the letter "W". The "M" and "W" are "ceiling designators".

A *ceiling designator* always precedes the height of the ceiling layer. Table 2-2 lists and explains ceiling designators. At MDW the ceiling height was measured. JFK had an indefinite ceiling which was vertical visibility into a surface based total obscuration.

The sky cover and ceiling as determined from the ground represent as nearly as possible what the pilot should experience in flight. In other words, a pilot flying at or above the reported ceiling layer aloft should see less than half the surface below him. The pilot descending through a surface based total obscuration should first see the ground directly below him from the height reported as vertical visibility into the obscuration. However, because of the differing viewing points of the pilot and the observer, these surface reported values do not always exactly agree with what the pilot sees. Figure 2-6 illustrates the effect of an obscured sky on the vision from a descending aircraft.

TABLE 2-2. Ceiling designators

Coded	Meaning	Spoken
M	Measured. Heights determined by ceilometer, ceiling light, cloud detection radar, or by the unobscured portion of a landmark protruding into the ceiling layer. (Figure 2-5 illustrates the principle of the ceilometer.)	MEASURED CEILING
E	Estimated. Heights determined from pilot reports, balloons, or other measurements not meeting criteria for measured ceiling.	ESTIMATED CEILING
W	Indefinite. Vertical visibility into a surface based obstruction. Regardless of method of determination, vertical visibility is classified as an indefinite ceiling.	INDEFINITE CEILING

The letter "V" appended to the ceiling height indicates a variable ceiling. The range of variability is shown in remarks. Variable ceiling is reported only when it is critical to terminal operations. As an example,

M12V OVC and in remarks CIG10V13

means MEASURED CEILING ONE THOUSAND TWO HUNDRED VARIABLE OVERCAST, CEILING VARIABLE BETWEEN ONE

Figure 2-3. Summation of cloud cover in multiple layers.

Figure 2-4. Summation of cloud cover in multiple layers. Note that at the height of the upper layer, sky cover is reported as overcast even though the upper layer itself covers less than 1/2 of the sky (10 SCT M25 BKN 80 OVC).

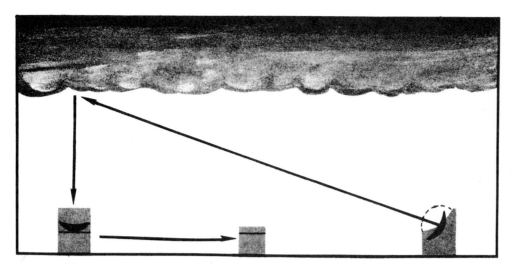

FIGURE 2-5. The rotating beam ceilometer. The projector beams a spot of modulated light on the cloud. The modulated light can be detected day or night. As the projector rotates, the spot moves along the cloud base. When the spot is directly over the detector, it excites a photoelectric cell measuring the angle of the light beam. Height of the cloud is then determined automatically by triangulation.

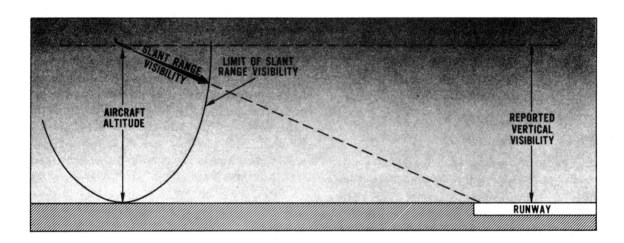

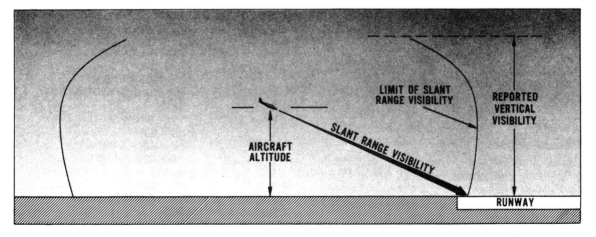

FIGURE 2-6. Vertical visibility is the altitude above the ground from which a pilot should first see the ground direclty below him (top). His real concern is slant range visibility which most often is less than vertical visibility. He usually must descend to a lower altitude (bottom) before he sees a representative surface and can fly by visual reference to the ground.

2-5

THOUSAND AND ONE THOUSAND THREE HUNDRED.

Now let's go back to our five reports and read them through sky and ceiling:

INK SA 1854 CLR	WINK, 1854 GREENWICH, CLEAR
BOI SA 1854 150 SCT	BOISE, 1854 GREENWICH, ONE FIVE THOUSAND SCATTERED
LAX SA 1852 7 SCT 250 SCT	LOS ANGELES, 1852 GREENWICH, SEVEN HUNDRED SCATTERED, TWO FIVE THOUSAND SCATTERED
MDW RS 1856 −X M7 OVC	CHICAGO MIDWAY, RECORD SPECIAL, 1856 GREENWICH, SKY PARTIALLY OBSCURED, MEASURED CEILING SEVEN HUNDRED OVERCAST
JFK RS 1853 W5 X	NEW YORK KENNEDY, RECORD SPECIAL, 1853 GREENWICH, INDEFINITE CEILING FIVE HUNDRED SKY OBSCURED

VISIBILITY

Prevailing visibility at the observation site immediately follows sky and ceiling in the report. Prevailing visibility is the greatest distance objects can be seen and identified through at least 180 degrees of the horizon. It is reported in statute miles and fractions.

Prevailing visibilities in the five reports are:

INK	VISIBILITY ONE FIVE
BOI	VISIBILITY THREE ZERO
LAX	VISIBILITY SIX
MDW	VISIBILITY ONE AND ONE-HALF
JFK	VISIBILITY ONE-QUARTER

When visibility is critical at an airport with a weather observing station and a control tower, both take visibility observations. When tower visibility is less than 4 miles, the lowest reported visibility of the 2 observations (surface, tower) is the prevailing visibility. The other is reported in remarks. Otherwise surface visibility is the reported prevailing visibility. Note that the report from JFK has a remark,

SFC VSBY1/2 meaning SURFACE VISIBILITY ONE-HALF.

The letter "V" suffixed to prevailing visibility denotes a variable visibility. The range of variability is shown in remarks. Variable visibility is reported only when critical to aircraft operations. As an example,

3/4V and in remarks VSBY1/2V1

means VISIBILITY THREE QUARTERS VARIABLE...VISIBILITY VARIABLE BETWEEN ONE-HALF AND ONE.

Visibility in some directions may differ significantly from prevailing visibility. These significant differences are reported in remarks. For example, prevailing visibility is reported as 1 1/2 miles with a remark,

VSBY NE21/2SW3/4

which means visibility to the northeast is 2-1/2 miles and to the southwest, it is 3/4 of a mile.

WEATHER AND OBSTRUCTIONS TO VISION

Weather and obstructions to vision when occurring at the station at observation time are reported immediately following visibility. If observed at a distance from the station, they are reported in remarks.

The term *weather* as used for this element refers only to those items in table 2-3 rather than to the more general meaning of all atmospheric phenomena. Weather includes all forms of precipitation plus thunderstorm, tornado, funnel cloud and waterspout.

TABLE 2-3. Weather symbols and meanings

Coded	Spoken
Tornado	TORNADO
Funnel Cloud	FUNNEL CLOUD
Waterspout	WATERSPOUT
T	THUNDERSTORM
T+	SEVERE THUNDERSTORM
R	RAIN
RW	RAIN SHOWER
L	DRIZZLE
ZR	FREEZING RAIN
ZL	FREEZING DRIZZLE
A	HAIL
IP	ICE PELLETS
IPW	ICE PELLET SHOWER
S	SNOW
SW	SNOW SHOWER
SP	SNOW PELLETS
SG	SNOW GRAINS
IC	ICE CRYSTALS

Precipitation is reported in one of three intensities. The intensity symbol follows the weather symbol with meanings as follows:

    Light   −
    Moderate (no sign)
    Heavy   +

No intensity is reported for hail (A) or ice crystals (IC).

A thunderstorm is reported as "T" and a severe thunderstorm as "T+". A *severe thunderstorm* is one in which surface wind is 50 knots or greater and/or surface hail is 3/4 inch or more in diameter.

Obstructions to vision include the phenomena listed in table 2-4. No intensities are reported for obstructions to vision.

TABLE 2-4. Obstructions to vision—symbols and meanings

Coded	Spoken
BD	BLOWING DUST
BN	BLOWING SAND
BS	BLOWING SNOW
BY	BLOWING SPRAY
D	DUST
F	FOG
GF	GROUND FOG
H	HAZE
IF	ICE FOG
K	SMOKE

Now referring back to our initial five reports, INK and BOI report no weather or obstructions to vision and no entries appear in the reports. Note at this point that by definition, obstructions to vision are only reported for visibilities of 6 miles or less whereas weather symbols will be used regardless of visibility. LAX reports two obstructions to vision, haze and smoke. MDW reports heavy rain as weather and fog as an obstruction to vision. JFK reports fog; is this weather or obstruction to vision?

There are two types of remarks concerning either a surface based obscuration or an obscuring phenomenon aloft. These remarks are discussed here.

When obscuring phenomenon is surface based and partially obscures the sky, a remark reports tenths of sky hidden. For example,

    K6

means 6/10 of the sky is hidden by smoke. Now look at the report from MDW; how much of the sky is hidden and by what obscuring phenomenon?

Note the remark

    RF2

which means 2/10 of the sky is hidden by rain and fog.

A layer of obscuring phenomenon aloft is reported in the sky and ceiling portion the same as a layer of cloud cover. A remark identifies the layer as an obscuring phenomenon. For example,

    20 −BKN and a remark K20 −BKN

means a thin broken layer of smoke based at 2,000 feet above the surface.

SEA LEVEL PRESSURE

Sea level pressure is separated from the preceding elements by a space. It is transmitted in record hourly reports only. It is in three digits to the nearest tenth of a millibar with the decimal point omitted. Sea level pressure usually is greater than 960.0 millibars and less than 1050.0 millibars. The first 9 or 10 is omitted. To decode, prefix a 9 or 10 whichever brings it closer to 1000.0 millibars. Again going back to our five reports, sea level pressures are:

INK	1010.6 millibars
BOI	1018.1 millibars
LAX	1012.9 millibars
MDW	999.0 millibars
JFK	1018.0 millibars

TEMPERATURE AND DEW POINT

Temperature and dew point are in whole degrees Fahrenheit. They are separated from sea level pressure by a slash (/). If sea level pressure is not transmitted, temperature is separated from preceding elements by a space. Temperature and dew point are separated also by a slash. A minus sign precedes a temperature or dew point when below zero (0) degree F. From our five reports, we have:

INK...77/63	WINK...TEMPERATURE SEVEN SEVEN, DEW POINT SIX THREE
BOI...62/42	BOISE...TEMPERATURE SIX TWO, DEW POINT FOUR TWO
LAX...60/59	LOS ANGELES...TEMPERATURE SIX ZERO, DEW POINT FIVE NINER
MDW...63/61	CHICAGO MIDWAY...TEMPERATURE SIX THREE DEW POINT SIX ONE
JFK...68/64	NEW YORK KENNEDY...TEMPERATURE SIX EIGHT, DEW POINT SIX FOUR

Additional examples with minus values:

CAR...−4/−16 CARIBOU ME...TEMPERATURE MINUS FOUR, DEW POINT MINUS ONE SIX

FAR... 6/−8 FARGO ND...TEMPERATURE SIX, DEW POINT MINUS EIGHT

WIND

Wind follows dew point and is separated from it by a slash. Average one minute direction and speed are in four digits. The first two digits are direction *from* which the wind is blowing. It is in tens of degrees referenced to true north*, i.e., 01 is 10 degrees; 21 is 210 degrees; 36 is 360 degrees or north. The second two digits are speed in knots. A calm wind is reported as 0000.

If wind speed is 100 knots or greater, 50 is added to the direction code and the hundreds digit of speed is omitted. Example,

5908

means 090 degrees (09 + 50 = 59) at 108 knots.

A *gust* is a variation in wind speed of at least 10 knots between peaks and lulls. A *squall* is a sudden increase in speed of at least 15 knots to a sustained speed of 20 knots or more lasting for at least one minute. Gusts or squalls are reported by the letter "G" or "Q" respectively following the average one-minute speed and followed by the peak speed in knots. For example,

1522Q37

means wind 150 degrees at 22 knots with peak speed in squalls to 37 knots.

Winds decoded from our five reports are

INK	WIND ONE ONE ZERO DEGREES AT ONE TWO PEAK GUSTS ONE EIGHT
BOI	WIND ONE THREE ZERO DEGREES AT FOUR
LAX	WIND TWO FIVE ZERO DEGREES AT FOUR
MDW	WIND THREE TWO ZERO DEGREES AT FIVE
JFK	WIND ONE EIGHT ZERO DEGREES AT FOUR

When any part of the wind report is *estimated* (direction, speed, peak speed in gusts or squalls), the letter "E" precedes the wind group. Example,

---
*Wind direction for the local station is *broadcast* in degrees magnetic.

E1522G28

is decoded WIND ONE FIVE ZERO DEGREES ESTIMATED TWO TWO PEAK GUSTS ESTIMATED TWO EIGHT.

A few stations do not transmit sea level pressure, temperature and dew point; and these elements usually are not included in a special. When the elements are not transmitted, the wind group is separated from the preceding element by a space; i.e.,

CSM SP W5 X 2F 1705/990

is a special from Clinton-Sherman OK (CSM) *not* transmitting sea level pressure, temperature or dew point.

ALTIMETER SETTING

Altimeter setting follows the wind group and is separated from it by a slash. Normal range of altimeter settings is from 28.00 inches to 31.00 inches of mercury. The last three digits are transmitted with the decimal point omitted. To decode, prefix to the coded value either a 2 or a 3 whichever brings it closer to 30.00 inches. Examples.

996 means ALTIMETER TWO NINER NINER SIX, (29.96 inches)

013 means ALTIMETER THREE ZERO ONE THREE (30.13 inches)

REMARKS

Remarks, if any, follow altimeter setting separated from it by a slash. Certain remarks should be reported routinely and others the observer may include when considered significant to aviation. Often, some of the most important information in an observation may be the remarks portion.

Runway Visibility and Runway Visual Range

The first remark, when transmitted, should be runway visibility or runway visual range. Figure 2-7 illustrates the difference. The terms are defined as follows:

Runway visibility—the visibility from a particular location along an identified runway, usually determined by a transmissometer instrument. It is in miles and fractions of a mile. Figure 2-8 diagrams the principle of the transmissometer.

Runway visual range—the maximum horizontal distance down a specified instrument runway at which a pilot can see and identify standard high intensity runway lights. It is always determined using a transmissometer and is reported in hundreds of feet.

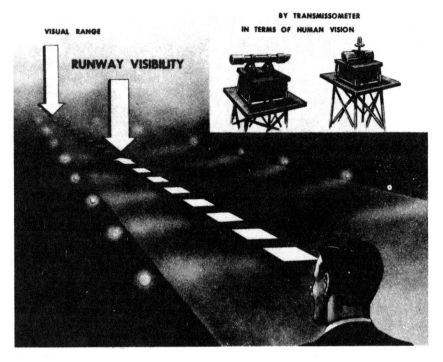

FIGURE 2-7. Difference between *runway visibility* and *runway visual range*. Runway visibility is the distance down the runway the pilot can see unlighted objects or unfocused lights of moderate intensity. Runway visual range is the distance he can see high intensity runway lights. Visual range usually is greater than visibility because the high intensity lights penetrate farther into the obscuring phenomena.

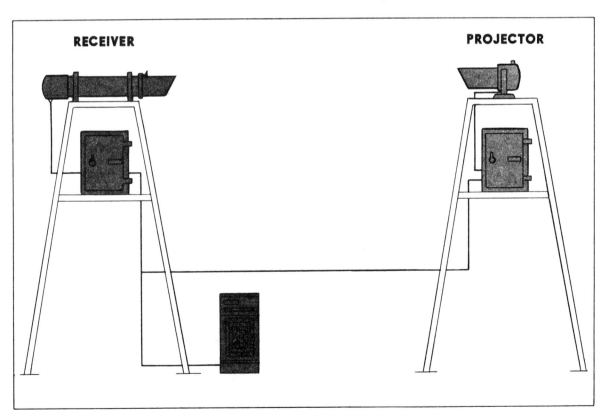

FIGURE 2-8. The transmissometer. The projector beams light toward the receiver. Obscuring phenomena in the path of the beam absorbs some of the light. A photoelectric cell in the receiver measures the amount of light penetrating through the obscuring phenomena. The amount received is converted into visibility.

2-9

The report consists of a runway designator and the contraction "VV" or "VR" followed by the appropriate visibility or visual range. Both the VV and the VR report are for a 10 minute period preceding observation time. The remark usually reports the 10 minute extremes separated by the letter "V". However, if the visual range or visibility has not changed significantly during the 10 minutes, a single value is sent indicating that the value has remained constant.

The following examples show several reports and their decoding:

R36VV11/2 RUNWAY THREE SIX, VISIBILITY VALUE ONE AND ONE-HALF. (Visibility remained constant during the 10 minutes period.)

R05LVV1V2 RUNWAY FIVE LEFT, VISIBILITY VALUE VARIABLE BETWEEN ONE AND TWO.

R18VR20V30 RUNWAY ONE EIGHT, VISUAL RANGE VARIABLE BETWEN TWO THOUSAND FEET AND THREE THOUSAND FEET.

R26RVR24 RUNWAY TWO SIX RIGHT, VISUAL RANGE TWO THOUSAND FOUR HUNDRED FEET. (Visual range remained constant during the 10 minute period.)

Runway visual range in excess of 6,000 feet is written 60+. VR less than the minimum value that can be observed by the instrument is encoded as the minimum suffixed by a minus sign. For example:

R36LVR10-V25

is decoded RUNWAY THREE SIX LEFT, VISUAL RANGE VARIABLE FROM LESS THAN ONE THOUSAND FEET TO TWO THOUSAND FIVE HUNDRED FEET.

Heights of Bases and Tops of Sky Cover Layers

Bases and tops of clouds or obscuring phenomena may be reported. These remarks originate from pilots (i.e., PIREPs). Therefore, heights are MSL. Examples,

/UA.../SK BKN 50

means top of broken layer 5,000 feet (MSL).

/UA.../SK OVC 30/60 OVC

means top of lower overcast 3,000 feet, base of higher overcast 6,000 feet (MSL).

PIREPs which are more than 15 minutes old are omitted unless it is considered to be operationally significant.

Clarification of Coded Data

Following, by category, are coded remarks clarifying or expanding on coded elements:

Sky and Ceiling

Coded Elements	Coded Remarks
FEW CU	Few cumulus clouds
HIR CLDS VSB	Higher clouds visible
BINOVC	Breaks in overcast
BRKS N	Breaks north
BKN V OVC	Broken layer variable to overcast
CIG 14V19	Ceiling variable between 1400 feet and 1900 feet
ACCAS ALQDS*	Altocumulus castellanus all quadrants
ACSL SW-NW*	Standing lenticular altocumulus southwest to northwest
ROTOR CLDS NW*	Rotor clouds northwest
VIRGA E-SE*	Virga (precipitation not reaching the ground) east through southeast
30 SCT V BKN	Scattered layer at 3000 feet variable to broken
SC BANK NW	Stratocumulus cloud bank northwest
TCU W*	Towering cumulus clouds west
CB N MOVG E*	Cumulonimbus north moving east
CBMAM OVHD-W*	Cumulonimbus mamma overhead to west
CONTRAILS N 420 MSL	Condensation trails north at 42,000 feet MSL
CLDS TPG MTNS SW	Clouds topping mountains southwest
RDGS OBSCD W-N	Ridges obscured west through north
CUFRA W APCHG STN	Cumulus fractus clouds west approaching station
LWR CLDS NE	Lower Clouds northeast

*These cloud types are highly significant and the observer should always report them. Figure 2-9 through 2-14 are photographs of these clouds and explains their significance. A pilot in flight should also report them when observed.

Obscuring Phenomena

Coded Elements	Coded Remarks
D5	Dust obscuring 5/10 of the sky
S7	Snow obscuring 7/10 of the sky
BS3	Blowing snow obscuring 3/10 of the sky

Coded Elements	Coded Remarks
FK4	Fog and smoke obscuring 4/10 of the sky
K20 SCT	Scattered layer of smoke aloft based at 2,000 feet above the surface
THN F NW	Thin fog northwest from reporting station)

Visibility (Statute Miles)

Coded Elements	Coded Remarks
VSBY S1W1/4	Visibility south 1, west 1/4
VSBY 1V3	Visibility variable between 1 and 3
TWR VSBY 3/4	Tower visibility 3/4
SFC VSBY 1/2	Surface visibility 1/2

Weather and Obstruction to Vision

Coded Elements	Coded Remarks
T W MOVG E FQT LTGCG	Thunderstorm west moving east, frequent lightning cloud to ground
RB30	Rain began 30 minutes after the hour
SB15E40	Snow began 15, ended 40 minutes after the hour
UNCONFIRMED TORNADO 15W OKC MOVG NE 2000	Unconfirmed tornado 15 (NM) west of Oklahoma City, moving northeast, sighted at 2000Z
T OVHD MOVG E	Thunderstorm overhead, moving east
OCNL DSNT LTG NW	Occassional distant lightning northwest
HLSTO 2	Hailstones 2 inches in diameter
INTMT R−	Intermittent light rain
DUST DEVILS NW	Dust devils northwest
OCNL RW	Occasional moderate rain shower
WET SNW	Wet Snow
SNOINCR 5	Snow increase 5 inches during past hour
R− OCNLY R+	Light rain occasionally heavy rain
RWU	Rain shower of unknown intensity
F DSIPTG	Fog dissipating
K DRFTG OVR FLD	Smoke drifting over field
KOCTY	Smoke over city
SHLW GFDEP 4	Shallow ground fog 4 feet deep
PATCHY GF S	patchy ground fog south

Wind

Coded Elements	Coded Remarks
WSHFT 30	Wind shifted at 30 minutes past the hour
WND 27V33	Wind variable between 270 degrees and 330 degrees
PK WND 3348/22	Peak wind within the past hour from 330 degrees at 48 knots occurred 22 minutes past the hour

Pressure

Coded Elements	Coded Remarks
PRESRR	Pressure rising rapidly
PRESFR	Pressure falling rapidly
LOWEST PRES 631 1745	Lowest pressure (sea level) 963.1 millibars at 1745 GMT
PRJMP 8/1012/18	Pressure jump (sudden increase) 0.08 inches began 1012 GMT, ended 1018 GMT

Freezing Level Data

Upper air (rawinsonde) observation stations append in remarks *freezing level data*. The coded remark is appended to the first record report transmitted after the information becomes available. Code for the remark is as follows:

RADAT UU (D) ($h_p h_p h_p$) ($h_p h_p h_p$) (/n)

(a) RADAT—a contraction identifying the remark as "freezing level data.".

(b) UU—relative humidity at the freezing level in percent. When more than one level is sent, "UU" is the highest relative humidity observed at any of the levels transmitted.

(c) (D)—a coded letter "L", "M", or "H" to indicate that relative humidity is for the "lowest", "middle", or "highest" level coded. This letter is omitted when only one level is sent.

(d) ($h_p h_p h_p$)—a height in hundreds of feet above MSL at which the upper air sounding crossed the zero (0) degree Celsius isotherm. No more than three levels are coded. If the sounding crosses the zero (0) degree Celsius isotherm more than three times, the levels coded are the lowest and the top two levels.

FIGURE 2-9. Towering Cumulus (TCU). The most direct significance of this cloud is that the atmosphere in the lower altitudes is unstable and conducive to turbulence.

FIGURE 2-10. Cumulonimbus (CB). The anvil portion of a CB is composed of ice crystals. The CB or thunderstorm cloud contains most types of aviation weather hazards; particularly turbulence, icing, hail, and low level wind shear (LLWS).

FIGURE 2-11. Cumulonimbus Mamma (CBMAM). This characteristic cloud can result from violent up and down currents and is often associated with severe weather. It indicates possible severe or greater turbulence.

FIGURE 2-12. Altocumulus Castellanus (ACCAS). ACCAS indicates unstable conditions aloft, but not necessarily below the base of the cloud. Note in this picture a surface based inversion shown by the trapped smoke, indicating stable conditions at the surface. Thus, rising air causing the ACCAS is originating somewhere above the surface based inversion. Compare with towering cumulus, a cloud representing unstable air and turbulence from the surface upward.

FIGURE 2-13. Virga. Virga is precipitation falling from a cloud but evaporating before reaching the ground. Virga results when air below the cloud is very dry and is common in the western part of the country. **Virga associated with showers suggests strong downdrafts with possible moderate or greater turbulence.**

FIGURE 2-14. Standing Lenticular Altocumulus (ACSL). **These clouds are characteristic of the standing or mountain wave.** A similar cloud is the Standing Lenticular Cirrocumulus (CCSL). CCSL are whiter and at higher altitude. Both are indicative of possible severe or greater turbulence.

(e) (/n)—indicator to show the number of crossings of the zero (0) degree Celsius isotherm, other than those coded. The indicator is omitted when all levels are coded.

Examples:

RADAT 87045	Relative humidity 87%, only crossing of zero (0) degrees C isotherm was 4,500 feet MSL.
RADAT 87L024105	Relative humidity 87% at the lowest (L) crossing. Two crossings occurred at 2,400 and 10,500 feet MSL.*
RADAT 84M019045051/1	Relative humidity 84% at the middle (M) crossing of the three coded crossings. Coded crossings were at 1,900, 4,500 and 5,100 feet. The 84% humidity was at 4,500 feet MSL. "/1" indicates one additional crossing and it was between 1,900 and 4,500 feet.
RADAT MISG	The sounding terminated below the first crossing of the zero (0) degree C isotherm—temperatures were all above freezing.
RADAT ZERO	The entire sounding was below zero (0) degree C.

*Temperature was below zero (0) degree C below 2,400 feet MSL; above zero (0) degrees C between 2,400 feet MSL and 10,500 feet MSL; and below zero (0) degrees C above 10,500 feet MSL.

Icing Data

When the rawinsonde observer determines definitely that icing is occurring on his instruments, he enters the data in the following code:

RAICG HHMSL (SNW)

(a) RAICG—indicates icing data follows.
(b) HH—height in hundreds of feet at which icing occurred. "MSL" is always appended to the height.
(c) (SNW)—used to indicate that snow is causing a reduced balloon ascension rate. (Omitted otherwise.)

Examples:

RAICG 12MSL—Icing at 1,200 feet MSL.

RAICG 24MSL SNW—Icing at 2,400 feet MSL in snow.

Other Information

A group or groups of numerically coded data may appear in remarks. These data are primarily of concern to the meteorologist and are not discussed here.

A printed arrow marks the end of weather information and signifies that the rest of the report is notice(s) to airmen (NOTAM). The NOTAM code is explained in the AIRMAN'S INFORMATION MANUAL.

REPORT IDENTIFIERS

A heading begins the record hourly collective on the local circuit identifying the type of message, the circuit number, data and time of observations making up the collective reports. For example,

SA21 271900

means surface aviation reports (SA), 21 is the circuit number, 27 is the day of the month, and the observations were made at 1900 GMT.

A slightly different heading begins each relay. It identifies the location of reporting stations by states and indicates the time the relay began. For example,

MO20192

means the relay is from Missouri, day of the month is the 20th (20), time of observations is 1900 GMT (19), and the relay began 12 minutes past the hour (12).

Relay designators other than states are INTERMEDIATE EAST, FAR EAST, NEAR NORTH, etc. The relay collectives are assembled by a centralized computer and are unique to each circuit.

Individual reports must each convey the time and type of report. Following are examples:

Example 1

INK SA 1100.....

indicates a relayed report from Wink, Texas for 1100 GMT (all times transmitted in teletypewriter reports are GMT). The "SA" signifies a record hourly.

Example 2

INK SA COR 1100.....

signifies a correction to the 1100 GMT record hourly report as originally transmitted. The correction may transmit the complete corrected report or it may contain only the corrected element or elements.

Example 3

INK SP 2315.....

indicates a special report of an observation taken at 2315 GMT to report a significant change in weather.

Example 4

INK SP COR 2315.....

indicates a correction to the above special report.

Example 5

BVO SW 1130.....

2-15

indicates a Supplemental Aviation Weather Reporting Station (SAWRS) report by the contraction "SW". SAWRS reports are unscheduled and are made by non-Government observers at airports not served by a regularly reporting weather station. Observations are taken during commercial aircraft operations. Type and time are transmitted. This report was from Bartlesville OK at 1130 GMT.

READING THE SURFACE AVIATION WEATHER REPORT

Now that we have studied the individual elements and their decoding, let's read completely each of the five reports. Capitalized phrases are those elements which *normally* are broadcast by the station at or near the airport where the observation was made:

INK SA 1854 CLR 15 106/77/63/1112G18/000

WINK, WINK, 1854 GREENWICH, CLEAR, VISIBILITY ONE FIVE, pressure 1010.6 millibars, TEMPERATURE SEVEN SEVEN, DEW POINT SIX THREE, WIND ONE ONE ZERO DEGREES AT ONE TWO PEAK GUSTS ONE EIGHT, ALTIMETER THREE ZERO ZERO ZERO.

BOI SA 1854 150 SCT 30 181/62/42/1304/015

BOISE, BOISE, 1854 GREENWICH, ONE FIVE THOUSAND SCATTERED, VISIBILITY THREE ZERO, pressure 1018.1 millibars, TEMPERATURE SIX TWO, DEW POINT FOUR TWO, WIND ONE THREE ZERO DEGREES AT FOUR, ALTIMETER THREE ZERO ONE FIVE.

LAX SA 1852 7 SCT 250 SCT 6HK 129/60/59/2504/991

LOS ANGELES, LOS ANGELES, 1852 GREENWICH, SEVEN HUNDRED SCATTERED, TWO FIVE THOUSAND SCATTERED, VISIBILITY SIX, HAZE, SMOKE. pressure 1012.9 millibars, TEMPERATURE SIX ZERO, DEW POINT FIVE NINER, WIND TWO FIVE ZERO DEGREES AT FOUR, ALTIMETER TWO NINER NINER ONE.

MDW RS 1856 −X M7 OVC 1 1/2 R+F 990/63/61/3205/980/RF2 RB12

CHICAGO, CHICAGO MIDWAY, RECORD SPECIAL, 1856 GREENWICH, SKY PARTIALLY OBSCURED, MEASURED CEILING SEVEN HUNDRED OVERCAST, VISIBILITY ONE AND ONE-HALF, HEAVY RAIN, FOG, pressure 999.0 millibars, TEMPERATURE SIX THREE, DEW POINT SIX ONE, WIND THREE TWO ZERO DEGREES AT FIVE, ALTIMETER TWO NINER EIGHT ZERO, TWO TENTHS SKY OBSCURED BY RAIN AND FOG, rain began 12 minutes past the hour.

JFK RS 1853 W5 X 1/4F 180/68/64/1804/006/R04RVR22V30 SFC VSBY 1/2

NEW YORK, NEW YORK KENNEDY, RECORD SPECIAL, 1853 GREENWICH, INDEFINITE CEILING FIVE HUNDRED SKY OBSCURED, VISIBILITY ONE-QUARTER, FOG, pressure 1018.0 millibars, TEMPERATURE SIX EIGHT, DEW POINT SIX FOUR, WIND ONE EIGHT ZERO DEGREES AT FOUR, ALTIMETER THREE ZERO ZERO SIX, RUNWAY FOUR RIGHT VISUAL RANGE VARIABLE BETWEEN TWO THOUSAND TWO HUNDRED FEET AND THREE THOUSAND FEET, SURFACE VISIBILITY ONE HALF.

AUTOMATED SURFACE OBSERVATIONS

Automatic Meteorological Observing Station AMOS

The AMOS is a solid-state system capable of automatically observing temperature, dew point, wind direction and speed, pressure (altimeter setting), peak wind speed, and precipitation accumulation. The field sensors are tied in directly to the FAA observation network. It transmits a weather report whenever the station is polled by the circuits. At a staffed AMOS, the observer can manually enter additional information to give a more complete observation. Tables 2-1, 2-2, 2-3 and 2-4 are used to interpret sky condition and weather. Visibility is in statute miles. The intensity symbols for precipitation are light (−), moderate (no sign), and heavy (+).

Figure 2-15 is the breakdown of an unstaffed AMOS.

Example:

PGO AMOS 76/61/0308/007 PK WND 18 013
which decodes as follows:

Observation from Page OK, temperature 76 degrees F, dew point 61 degrees F, wind from 030 degrees at 8 knots, altimeter setting 30.07 inches, peak wind since the last hourly observation 18 knots, 13 hundredths of an inch of liquid precipitation since last synoptic observation.

Figure 2-16 is the breakdown of a staffed AMOS.
Example:

GLS AMOS SA 1356 E12 BKN 150 OVC 7TRW-F 79/77/1103/004 PK WND 15 008/TB32 NW MOVG W

MDO AMOS 33/29/3606/975 PK WND 08 001

MDO	STATION IDENTIFICATION: (Middleton Island AK) Identifies report using FAA identifiers.
AMOS	AUTOMATIC STATION IDENTIFIER
33	TEMPERATURE: (33 degrees F.) Minus sign indicates sub-zero temperatures.
29	DEW POINT: (29 degrees F.) Minus sign indicates sub-zero temperatures.
3606	WIND: (360 degrees true at 6 knots. Direction is first two digits and is reported in tens of degrees. To decode, add a zero to first two digits. The last digits are speed; e.g., 2524=250 degrees at 24 knots.
975	ALTIMETER SETTING: (29.75 inches) The tens digit and decimal are omitted from report. To decode prefix a 2 to code if it begins with an 8 or 9. Otherwise prefix a 3; e.g., 982=29.82, 017=30.17.
PK WND 08	PEAK WIND SPEED: (8 knots) Reported speed is highest detected since last hourly observation.
001	PRECIPITATION ACCUMULATION: (0.01 inches. Amount of precipitation since last synoptic time (00, 06, 12, 1800 GMT).

FIGURE 2-15. Decoding observations from unstaffed AMOS stations.

SMP SP 0056 AMOS -X M20 BKN 7/8L-FK 046/66/65/2723/967 PK WND 36 027/VSBY S 1/4

SMP	STATION IDENTIFICATION: (Stampede Pass WA) Identifies report using FAA identifiers.
SP	TYPE OF REPORT: (Special) SA=Record
0056	TIME OF REPORT: GMT
AMOS	AUTOMATIC STATION IDENTIFIER
-X M20 BKN	SKY & CEILING: (partly obscured sky, ceiling measured 2,000 feet broken) Figures are height in 100s of feet above ground. Number preceding an X is vertical visibility into a total obscuration in 100s of feet. Symbol after height is amount of sky cover. Letter preceding height indicates method used to determine height.
7/8	PREVAILING VISIBILITY: (Seven-eighths statute miles)
L-FK	WEATHER & OBSTRUCTIONS TO VISION: (Light drizzle, Fog, & Smoke) Algebraic signs indicate intensity.
046	SEA-LEVEL PRESSURE: (1004.6 millibars) Only the tens, units, and tenths digits are reported.
66	TEMPERATURE: (66 degrees F.) Minus sign indicates sub-zero temperatures.
65	DEW POINT: (65 degrees F.) Minus sign indicates sub-zero temperatures.
2723	WIND: (270 degrees true at 23 knots) Direction is first two digits and is reported in tens of degrees. To decode, add a zero to first two digits. The last digits are speed; e.g., 2524=250 degrees at 24 knots.
967	ALTIMETER SETTING: (29.67 inches) The tens digit and decimal are omitted from report. To decode prefix a 2 to code if it begins with an 8 or 9. Otherwise prefix a 3; e.g., 982=29.82, 017=30.17.
PK WND 36	PEAK WIND SPEED: (36 knots) Reported speed is highest detected since last hourly observation.
027	PRECIPITATION ACCUMULATION: (0.27 inches.) Amount of precipitation since last synoptic time (00, 06, 12, 1800 GMT).
VSBY S 1/4	MISCELLANEOUS REMARKS & NOTAMS: (Visibility to south 1/4 mile) Remarks are given using contractions.

FIGURE 2-16. Decoding observations from staffed AMOS stations.

This Galveston TX observation contains similar information to the PGO example above but with the addition of type and time of observation, sky condition, visibility and weather, and remarks made by an observer.

Examples of a staffed AMOS observation:

GLS AMOS SA 1455 CLR 7 82/71/0311/010 PK WND 15 000 GLS AMOS SA 1755 11 SCT E100 OVC 7 80/73/0706/996 PK WND 13 028/TCU ALQDS RWU SW
GLS AMOS RS 1555 7 SCT 15 SCT E60 OVC 7R− 78/75/0506/006 PK WND 12 008

Automatic Observing Station (AUTOB)

The AUTOB is an AMOS with added capability to *automatically* report sky conditions, visibility and precipitation occurrence. AUTOB is polled at 20 minute intervals. The upper limit of cloud amount and height measurements is 6,000 feet AGL. Visibility in statute miles is determined by a backscatter sensor with reportable categories of 0 to 8 (see table 2-5). If a visibility report consisting of 3 values is encountered, it is decoded as shown in the following example:

"VB786", 7 = present visibility, 8 = maximum visibility during past 10 minutes, and 6 = minimum visibility during past 10 minutes.

TABLE 2-5. Reportable visibility categories.

Index of vis. (stat. mi.)	When vis. is:	Index of vis. (stat. mi.)	When vis is:
0	less than 15/16	5	4 1/2 - 5 1/2
1	1 - 1 7/8	6	5 1/2 - 6 1/2
2	2 - 2 7/8	7	6 1/2 - 7 1/2
3	3 - 3 1/2	8	above 7 1/2
4	3 1/2 - 4 1/2		

AUTOB may indicate no cloud layers in either a clear situation or during an inability to penetrate a surface-based obscuration. To distinguish the two, the following rules apply. If the visibility is less than 2 miles, either a partial obscuration "−X" or indefinite obscuration "WX" is reported. A "−X" implies some cloud returns and a "WX" implies no cloud returns. A vertical visibility value for "WX" is not measured. When visibility is 2 miles or greater and no cloud returns are detected a "CLR BLO 60" is used which indicates a clear sky below 6,000 feet. "E" is the ceiling designator. A maximum of 3 (lowest) cloud layers will be reported. Figure 2-17 is the breakdown of an AUTOB message.

Example:

DRT AUTOB 25 SCT E40 OVC BV5 P 58/52/ 1412/995 PK WND 16 004

This Del Rio TX observation indicates: sky condition of two thousand five hundred feet scattered, ceiling four thousand feet overcast, surface visibility between 4 1/2 and 5 1/2 statute miles, precipitation has occurred within 10 minutes of the observation, temperature 58 degrees F, dew point 52 degrees F, wind from 140 degrees at 12 knots, altimeter setting 29.95 inches, peak wind since the last hourly observation 16 knots, four hundredths (0.04 inches) of liquid precipitation since the last synoptic observation.

Examples of an AUTOB message:

DRT AUTOB CLR BLO 60 BV8 75/65/0905/ 991 PK WND 08 000
DRT AUTOB 25 SCT E31 OVC BV8 83/71/ 1408/989 PK WND 18 120
DRT AUTOB 5 SCT E10 OVC BV4 P 75/74/ 1306/978 PK WND 07 001
DRT AUTOB −X E5 BKN BV1 68/67/1805/ 992 PK WND 09 000
DRT AUTOB WX BV0 70/70/1401/995 PK WND 02 000
DRT AUTOB −X 8 SCT E13 OVC BV1 73/ 69/1307/000 PK WND 10 000 HIR CLDS DETECTED

A remark "HIR CLDS DETECTED" is included if clouds are detected above an overcast, with the higher clouds "HIR CLDS" being less than 6,000 feet. For example:

E30 OVC BV8 65/58/3606/988 PK WND 10 000 HIR CLDS DETECTED

means higher clouds are being detected above the overcast, but are less than 6,000 feet. Note that an AUTOB makes no distinction between a thin and opaque cloud layer. "E30 OVC" may be a thin overcast, but is reported as a ceiling.

Remote Automatic Meteorological Observing System (RAMOS)

The breakdown of a RAMOS message is shown in figure 2-18. Note the similarity to the unstaffed AMOS observation except that a 3-hour pressure change, maximum/minimum temperature, and 24-hour precipitation accumulation is also included at designated times.

2-18

ENV AUTOB E25 BKN BV7 P 33/29/3606/975 PK WND 08 001

ENV	STATION IDENTIFICATION: (Wendover UT) Identifies report using FAA identifiers.
AUTOB	AUTOMATIC STATION IDENTIFIER
E25 BKN	SKY & CEILING: (Estimated 2,500 feet broken) Figures are height in 100s of feet above ground. Contraction after height is amount of sky cover. Letter preceding height indicates ceiling. WX reported if visibility is less than 2 miles and no clouds are detected. NO CLOUDS WILL BE REPORTED ABOVE 6,000 FEET.
BV7	PRESENT VISIBILITY: Reported in whole statute miles from 0 to 8. Visibility is averaged over a 10 minute time period.
P	PRECIPITATION OCCURRENCE: (P=precipitation in past 10 minutes).
33	TEMPERATURE: (33 degrees F.) Minus sign indicates sub-zero temperatures.
29	DEW POINT: (29 degrees F.) Minus sign indicates sub-zero temperatures.
3606	WIND: (360 degrees true at 6 knots. Direction is first two digits and is reported in tens of degrees. To decode, add a zero to first two digits. The last digits are speed; e.g., 2524=250 degrees at 24 knots.
975	ALTIMETER SETTING: (29.75 inches) The tens digit and decimal are omitted from report. To decode prefix a 2 to code if it begins with an 8 or 9. Otherwise prefix a 3; e.g., 982=29.82, 017=30.17.
PK WND 08	PEAK WIND SPEED: (8 knots) Reported speed is highest detected since last hourly observation.
001	PRECIPITATION ACCUMULATION: (0.01 inches) Amount of precipitation since last synoptic time (00, 06, 12, 1800 GMT).

NOTE: If no clouds are detected below 6,000 feet and the visibility is greater than 2 miles, the reported sky condition will be CLR BLO 60.

FIGURE 2-17. Decoding observations from AUTOB stations.

```
P67   RAMOS   SA   2356   046/66/65/2723/967 PK WND 36
0002 027 83 20043
```

P67	STATION IDENTIFICATION: (Lidgerwood ND) Identifies report using FAA identifiers.
RAMOS	AUTOMATIC STATION IDENTIFIER
SA	TYPE OF REPORT: (Record) SP=Special
2356	TIME OF REPORT: GMT
046	SEA-LEVEL PRESSURE: (1004.6 millibars) Only the tens, units, and tenths digits are reported.
66	TEMPERATURE: (66 degrees F.) Minus sign indicates sub-zero temperatures.
65	DEW POINT: (65 degrees F.) Minus sign indicates sub-zero temperatures.
2723	WIND: (270 degrees true at 23 knots) Direction is first two digits and is reported in tens of degrees. To decode, add a zero to first two digits. The last digits are speed; e.g., 2524=250 degrees at 24 knots.
967	ALTIMETER SETTING: (29.67 inches) The tens digit and decimal are omitted from report. To decode prefix a 2 to code if it begins with an 8 or 9. Otherwise prefix a 3; e.g., 982=29.82, 017=30.17.
PK WND 36	PEAK WIND SPEED: (36 knots) Reported speed is highest detected since last hourly observation.
0002	THREE-HOUR PRESSURE CHANGE: (Rising then falling, 0.02 millibars higher now than three hours ago.) ‡
027	PRECIPITATION ACCUMULATION: (0.27 inches) Amount of precipitation since last synoptic time (00, 06, 12, 1800 GMT).
83	TEMPERATURE (MAX OR MIN) MAX at 00 & 06Z, MIN at 12 & 18Z
20043	PRECIPITATION ACCUMULATION IN PAST 24 HOURS: (00.43 inches); first digit (2) is the group identifier

---

‡ First digit in code is barometer tendency (see figure 5-6).

```
0 = ∧        5 = ∨
1 = ⌐        6 = ⌊
2 = /        7 = \
3 = √        8 = ∧
4 = ∧,-,∨
```

FIGURE 2-18. Decoding observations from RAMOS stations.

Section 3
PILOT AND RADAR REPORTS AND SATELLITE PICTURES

The preceding section explained the decoding of surface aviation weather reports. However, these spot reports only sample the total weather picture. Pilot and radar reports along with satellite pictures help fill the gaps between stations.

PILOT WEATHER REPORTS (PIREPS)

No observation is more timely than the one you make from your cockpit. In fact, aircraft in flight are the only means of directly observing cloud tops, icing and turbulence. Your fellow pilots welcome your PIREP as well as do the briefer and forecaster. Help yourself and the aviation weather service by sending pilot reports!

A PIREP is usually transmitted in a prescribed format (see figure 3-1). The letters "UA" identify the message as a pilot report. The letters "UUA" identify an urgent PIREP. Required elements for all PIREPs are message type, location, time, flight level, type of aircraft and at least one weather element encountered. When not required, elements without reported data are omitted. All altitude references are MSL unless otherwise noted, distances are in nautical miles and time is in GMT.

A PIREP is usually transmitted as part of a **group of PIREPs** collected by state or as a **remark appended** to a surface aviation weather report. The phenomenon is coded in contractions and symbols. For example (referring to Figure 3-1 as a guide):

UA /OV MRB-PIT/TM 1600/FL 100 /TP BE55 /SK 024 BKN 032/042 BKN-OVC /TA -12/IC LGT-MDT RIME 055-080/RM WND COMP HEAD 020 MH310 TAS 180. *

The PIREP decodes as follows:

Pilot report, Martinsburg to Pittsburgh at 1600Z at 10,000 feet. Type of aircraft is a Beechcraft Baron. First cloud layer has base at 2,400 feet broken top 3,200 feet. Second cloud layer base is 4,200 feet broken occasionally overcast with no tops reported. Outside air temperature is −12 degree Celsius. Light to moderate rime icing reported between 5,500-8,000 feet. Headwind component is 20 knots. Magnetic heading is 310 degrees and true air speed is 180 knots.

---

* NOTE that all heights are referenced to MSL.

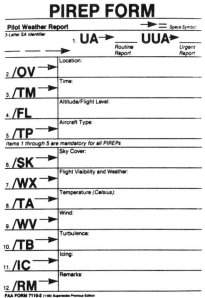

FIGURE 3-1. Pilot Reports Format.

3-1

The following example appended to an aviation weather report,

DSM SA 1755 M8 OVC 3R-F 132/45/44/3213/992/UA /OV DSM 320012/TM 1735/FL UKN /TP UKN /SK OVC 065/080 OVC 140. *

is decoded "...pilot report 12 nautical miles on 320 degrees radial from Des Moines VOR, at 1735 GMT, flight level and type of aircraft are unknown, top of the lower overcast 6,500 feet; base of a second overcast layer at 8,000 feet with top at 14,000 feet."

Note that PIREPs adhere to format shown in figure 3-1.

UA /OV OKC 063064/TM 1522/FL080 TP C172/TA -04 /WV 245040 /TB LGT /RM IN CLR *

PIREP decodes as follows:

Pilot report, 64 nautical miles on 63 degree radial from Oklahoma City VOR at 1522 GMT at flight altitude 8,000 feet. Type aircraft is a Cessna 172. Outside air temperature is minus 4 degrees Celsius, wind is 245 degrees at 40 knots, light turbulence and clear skies.

Most contractions in PIREP messages are self-explanatory. Icing and turbulence reports state intensities using standard terminology when possible. Intensity tables for turbulence and icing are in section 14. If a pilot's description of an icing or turbulence encounter cannot readily be translated into standard terminology, the pilot's description is transmitted verbatim.

To lessen the chance of misinterpretation by others, you are urged to report icing and turbulence in standard terminology (intensity tables for turbulence and icing, section 14). This PIREP stated,...PRETTY ROUGH AT 6,500, SMOOTH AT 8,500 PA24... Would a report of "light", "moderate", or "severe" turbulence at 6,500 have meant more to you?

Pilot reports of individual cloud layers, bases and tops, are usually in symbols and are often appended to surface aviation weather reports. Height of cloud base precedes the sky cover symbol and top follows the symbol. For example, 038 BKN 070 means base of broken layer at 3,800 feet and top 7,000 feet (all MSL).

Outside air temperature is given in 2 digits, degrees Celsius, with negative values preceded by a hyphen. Wind is given as six digits with the first 3 digits being direction and the last 3 digits being speed in knots.

The following excerpts may further assist you in reading transmitted pilot weather reports:

.../RM DURGD OAOI 150 01 080...*

means "...during descent on and off instruments from 15,000 feet; on instruments from 8,000 feet...

...FL100.../TA-02/WV 250015*

is decoded "...at 10,000 feet, temperature -2 degrees C, wind 250 degrees at 15 knots..."

...FL060/TP C-172/SK INTMTLY BL/TB MDT...*

states "...at 6000 feet, Cessna 172, intermittently between layers (contraction BL); moderate turbulence..."

UA /OV ABQ/TM 1845.. TIJERAS PASS CLOSED DUE TO FOG AND LOW CLDS. UNABLE VFR RTNG ABQ.*

is self-explanatory. Information of this type is helpful to others planning VFR flight in the area.

UA /OV/ TOL/TM 2200/FL 310 /TP B707 /TB MDT CAT 350-390*

means "...over Toledo at 2200 GMT and flight level 31,000, a Boeing 707 reported moderate clear air turbulence from 35,000 to 39,000.

Pilot reports of a non-meteorological nature sometimes help air traffic controllers. This "plain language" report stated:

"...3N PNS LRG FLOCK OF GOOSEY LOOKING BIRDS HDG GNLY NORTH MAY BE SEAGULLS FORMATION LOUSY COURSE ERRATIC..."

While in humorous vein, this PIREP alerted pilots and controllers to a bird hazard.

Your PIREP always helps someone else and becomes part of the aviation weather service. Please report anything you observe that may be of concern to other pilots.

RADAR WEATHER REPORTS (RAREPS)

Thunderstorms and general areas of precipitation can be observed by radar. Most radar stations report each hour at H+35 with intervening special reports as required. The report includes location of precipitation along with type, intensity, and intensity trend. Table 3-1 explains symbols denoting intensity and trend. Table 3-2 shows the order and content of a radar weather report.

---

*Note that all heights are referenced to MSL.

TABLE 3-1. Precipitation intensity and intensity trend

Intensity		Intensity Trend	
Symbol	Intensity	Symbol	Trend
−	Light	+	Increasing
(none)	Moderate		
+	Heavy	−	Decreasing
++	Very Heavy		
X	Intense	NC	No change
XX	Extreme		
U	Unknown	NEW	New echo

TABLE 3-2. Ordered content of a radar weather report

OKC 1934 LN 8TRW++/+ 86/40 164/60 199/115 15W L2425 MT570 AT 159/65 2 INCH HAIL RPRTD THIS CELL MO1 NO2 ON3 PM34 QM3 RL2 SL9

OKC 1934	LN	8	TRW++/+	86/40 164/60 199/115
a.	b.	c.	d.	e.

15W	L2425	MT 570 AT 159/65
f.	g.	h.

2 INCH HAIL RPRTD THIS CELL
i.

MO1 NO2 ON3 PM34 QM3 RL2 SL9
j.

See TABLE 7-1 for corresponding rainfall rates defining intensities. Note that intensity and intensity trend is not applicable to frozen precipitation.

Refer to TABLE 3-2:
a. Location identifier and time of radar observation (Oklahoma City RAREP at 1934 GMT in this example).
b. Echo pattern (line in this example)—
  Echo pattern or configuration may be a
  1. Line (LN)—a line of precipitation echoes at least 30 miles long, at least five times as long as it is wide and at least 30% coverage within the line.
  2. Fine Line (FINE LN)—a unique *clear air* echo (usually precipitation free and cloud free) in the form of a thin or fine line on the PPI scope. It represents a strong temperature/moisture boundary such as an advancing dry cold front.
  3. Area (AREA)—a group of echoes of similar type and not classified as a line.
  4. Spiral Band Area (SPRL BAND AREA)—an area of precipitation associated with hurricane that takes on a spiral band configuration around the center.
  5. Single Cell (CELL)—a single isolated precipitation not reaching the ground.
  6. Layer (LYR)—an elevated layer of stratform precipitation not reaching the ground.
c. Coverage in tenths (8/10 in this example).
d. Type, intensity, and intensity trend of weather (thunderstorm (T), very heavy rainshowers (RW++) and increasing in intensity (/+) in this example)—See TABLE 7-1 for weather symbols used except hail is reported as "A" in a RAREP. See TABLE 3-1 for intensity and intensity trend symbols.
e. Azimuth (reference true N) and range in nautical miles (NM) of points defining the echo pattern (86/40 164/60 199/115 in this example)—See following examples for elaboration of echo patterns.
f. Dimension of echo pattern (15 NM wide in this example)—Dimension of an echo pattern is given when azimuth and range define *only* the center line of the pattern. In this example, "15W" means the line has a total width of 15 NM, 7 1/2 miles either side of a center line drawn from the points given i.e. "D15" means a convective echo is 15 miles in diameter around a given center point.
g. Pattern movement (line moving *from* 240 degrees at 25 knots in this example)—may also show movement of individual storms or cells "C" and movement of an area "A".
h. Maximum top and location (57,000 feet MSL on radial 159 degrees at 65 NM in this example).
i. Remarks—self-explanatory using plain language contractions.
j. Digital section—used for preparing radar summary chart.

To assist you in interpreting RAREPs, four examples are decoded into plain language:

LZK 1133 AREA 4TRW+/+ 22/100 88/170 196/180 220/115 C2425 MT 310 AT 162/110
(Radial distance)

Little Rock AR radar weather observation at 1133 GMT. An area of echoes, four-tenths coverage, containing thunderstorms and heavy rainshowers, increasing in intensity.

Area is defined by points (referenced from LZK radar site) at 22 degrees, 100 NM (nautical miles); 88 degrees, 170 NM; 196 degrees, 180 NM and 220 degrees, 115 NM. (These points plotted on a map and connected with straight lines outline the area of echoes.

3-3

Maximum top (MT) is 31,000 feet MSL located at 162 degrees and 110 NM from LZK.

JAN 1935 SPL LN 10TRWX/NC 86/40 164/60 199/115 12W C2430 MT 440 AT 159/65 D10

Jackson MS special radar report at 1935 GMT. Line of echoes, ten-tenths coverage, thunderstorm, intense rainshowers, no change in intensity.
Center of the line extends from 86 degrees, 40 NM; 164 degrees, 60 NM to 199 degrees, 115 NM. The line is 12 NM wide (12W). (To display graphically, plot the center points on a map and connect the points with a straight line; since the thunderstorm line is 12 miles wide, it extends 6 miles either side of your plotted line.)
Thunderstorm cells are moving from 240 degrees at 30 knots.
Maximum top is 44,000 feet MSL centered at 159 degrees, 65 NM from JAN.
Diameter of this cell is 10 NM (D10).

MAF 1130 AREA 2S 27/80 90/125 196/50 268/100 A2410 MT U100

Midland TX radar weather report at 1130 GMT.
An area, two-tenths coverage, of snow (no intensity or trend is assigned for non-liquid precipitation) Area is bounded by points 27 degrees, 80 NM; 90 degrees, 125 NM; 196 degrees, 50 NM and 268 degrees, 100 NM.
Area movement is from 240 degrees at 10 knots.
Maximum tops are 10,000 feet MSL, tops are uniform (smooth). Note that these are precipitation tops and not cloud tops.

HDO 1132 AREA 2TRW++6R-/NC 67/130 308/45 105W C2240 MT 380 AT 66/54

Hondo TX radar weather report at 1132 GMT.
An area of echoes, total coverage eight-tenths, containing two-tenths coverage of thunderstorms with very heavy rainshowers and six-tenths coverage of light rain. No change in intensity. (Suggests thunderstorms embedded in an area of light rain.).
Although the pattern is an "area", only two points are given followed by "105W". This means the area lies 52 and 1/2 miles either side of the line defined by the two points - 67 degrees, 130 NM and 308 degrees, 45 NM.
Thunderstorm cells are moving from 220 degrees at 40 knots.
Maximum top is 38,000 feet at 66 degrees, 54 NM from HDO.

When a radar report is transmitted but doesn't contain any encoded weather observation, a contraction is sent which indicates the operational status of the radar. For example,

OKC 1135 PPINE means Oklahoma City OK radar at 1135 GMT detects no echoes.

TABLE 3-3 explains the contractions.

TABLE 3-3. Contractions of radar operational status.

Contraction	Operational status
PPINE	Equipment normal and operating in PPI (Plan Position Indicator) mode; no echoes observed.
PPIOM	Radar inoperative or out of service for preventative maintenance.
PPINA	Observations not available for reasons other than PPINE or PPIOM.
ROBEPS	Radar operating below performance standards.
ARNO	"A" scope or azimuth/range indicator inoperative.
RHINO	Radar cannot be operated in RHI (Range-height indicator) mode. Height data not available.

A radar weather report may contain remarks in addition to the coded observation. Certain types of severe storms may produce distinctive patterns on the radar scope. For example, a hook-shaped echo may be associated with a tornado. A line echo wave pattern (LEWP) in which one portion of a squall line bulges out ahead of the rest of the line may produce strong gusty winds at the bulge. A "vault" on the Range-Height Indicator scope may be associated with a severe thunderstorm producing large hail and strong gusty winds at the surface. If hail, strong winds, tornado activity, or other adverse weather is known to be associated with identified echoes on the radar scope, the location and type of phenomenon are included as a remark. Examples of remarks are, "HAIL REPORTED THIS CELL", "TORNADO ON GROUND AT 338/15" AND "HOOK ECHO 243/18". As far as indicating precipitation *not* reaching the ground, two contractions are aloft and mostly aloft respectively. That is, some or most of the precipitation is *not* reaching the ground. Bases of the precipitation will be given in hundred of feet MSL; example "PALF BASE 40" means part of the precipitation detected is evaporating at 4,000 feet MSL.

Radar weather reports also contain groups of digits, ie, MO1 NO2 ON3 PM34, etc., which are entered on a line following the RAREP. This digitized radar information (omitted from the foregoing examples) is used primarily by meteorologists and hydrologists for estimating amount of rainfall and in preparing the radar summary chart. However, this code is useful in determining more precisely where precipitation is occurring within an area and the intensity of the precipitation by using a proper grid overlay chart for the corresponding radar site. See Figure 3-2 for an example of a digital code plotted from the OKC RAREP in Table 3-2.

The digit assigned to a box represents encoded intensity levels of the precipitation as determined by a video integrator processor. See Table 7-1 for definitions of intensity levels 1-6. Thus, the term VIP LEVEL 1 simply means the precipitation intensity is weak or light, VIP LEVEL 2 is moderate, etc. Note that the *maximum* VIP LEVEL is encoded for any given box on the grid identified in the digital code. A box is identified by two letters, the first representing the row in which the box is found and the second letter representing the column. For example "MO1" identifies the box located in row M and column 0 as containing precipitation with a maximum VIP LEVEL of one (1). A code of "MO1324" indicates precipitation in four consecutive boxes in the same row. Working from left to right box MO = 1, box MP = 3, box MQ = 2 and box MR = 4.

When using hourly and special radar weather reports in preflight planning, note the location and

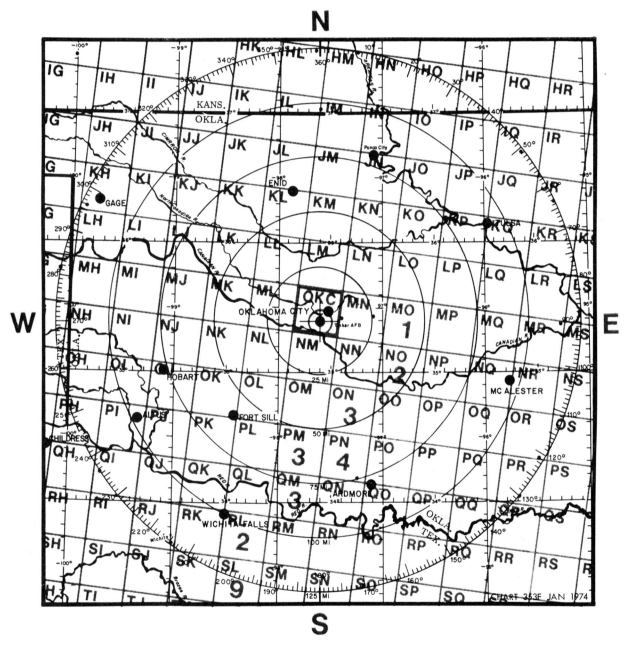

FIGURE 3-2. Digital Radar Report Plotted on a PPI Grid Overlay Chart. Data from Table 3-2. *Note*: See Table 7-1 for Intensity Level Codes 1 through 6.
The following VIP LEVEL codes are used for echoes beyond 125 nautical miles:
8 = Echoes of unknown intensity but believed to be severe from other reports.
9 = Echoes of unknown intensity but not believed to be severe.

3-5

Pilot and Radar Reports and Satellite Pictures

```
SDUS KNKA 112113
SDUS24 KWBC 112035
+ 51 041
                                  ........          .  25562213..   .
                                       .     .       . 24342  1.    .
                                       .     .       . 32543        .    .
                                       .     .       .  94411       .    .
                                       .     .       .  95421       .    .
                                       .     ..      ..  1522       .    .
                                       .     ........  ...1511...        .
                                       .                 .1511      .    .
                                       .                  .13       .
                                                           531         .  1 11
                                                          9531         .  2221
                                       .                  2512         ....221
                                       .                   33.         .  1221
                                                          11.2            1
                                                          11 .            ..2
                                                           1             21.
                                                                     .
                                       .    ....                     . 1
                                        ..MM.  ..                    . 1    1
                                          MMMMM    .                ..
                                           MMMM    .                ..
                                           MMMM     .               .
                                          MMMMM      .              .
                                          MMMMM       .             .
                                           MMMMM      .            .
                                                       ..
                                                        .....
```

FIGURE 3-3. Teletypewriter Plot of Echo Intensities for the South Central United States. *Note:* See Table 7-1 for Intensity Level Codes 1 through 6.
M = Missing.
8 = Echoes of unknown intensity (beyond 125 NM) but believed to be severe.
9 = Echoes of unknown intensity (beyond 125 NM) but not believed to be severe.

coverage of echoes, the type of weather reported, the intensity trend and especially the direction of movement.

A WORD OF CAUTION—remember that when National Weather Service radar detects objects in the atmosphere it only detects those of precipitation size or greater. It is *not* designed to detect ceilings and restrictions to visibility. An area may be blanketed with fog or low stratus but unless precipitation is also present the radar scope will be clear of echoes. Use radar reports along with PIREPs and aviation weather reports and forecasts.

RAREPs help you to plan ahead to avoid thunderstorm areas. Once airborne, however, you must depend on visual sighting or airborne radar to evade individual storms.

Another product provided for the use of the pilot in flight planning is the teletypewriter digital plot (see figure 3-3). The digital plot may be obtained through the request/reply circuit. The chart has digitized intensities plotted over a section map of several states. The numbers on the plot refer to RADAR intensity and represent the strongest return found in the area. From this plot the briefer/pilot may obtain the location of the strongest RADAR returns in the area of interest.

SATELLITE WEATHER PICTURES

Prior to the space age, weather observations were made only at distinct points within the atmosphere and complemented by pilot observations (PIREPs) of en route clouds and weather. These PIREPs give a "sense" of weather as viewed from above. However, with the advent of weather satellites a whole new dimension to weather observing and reporting has emerged. There are two types of weather satellites in use today: GOES (a geostationary satellite) and NOAA (a near polar orbiter satellite).

There are two U.S. GOES (Geostationary Operational Environmental Satellite) satellites used for picture taking. One stationed over the equator at 75 degrees west and the other at 135 degrees west. Together they cover North and South America and surrounding waters. They normally each transmit a picture of the earth, pole to pole each half hour. When disastrous weather threatens the U.S., the satellites can scan small areas rapidly so that we can receive a picture as often as every three minutes. Data from these rapid scans are used at national warning centers.

However, since the GOES satellite is stationary over the equator, the pictures poleward of about 50% degrees latitude become significantly distorted. Thus, another type of satellite is employed. The NOAA satellite is a near polar orbiter with an inclination angle (to the equator) of 98.6 degrees. In other words, this satellite orbits the earth on a track

FIGURE 3-4. GOES Visible Impagery.

which nearly crosses the North and South Poles. A high resolution picture is produced about 800 miles either side of it's track on the journey from pole to pole. The NOAA pictures are essential to weather personnel in Alaska and Canada.

NOTE: At the time of this revision, GOES East (75 degrees west) has become inoperative and GOES West has been positioned at 100 degrees west longitude in order to cover as much of the U.S. as possible. As a result, from the west coast out into the Pacific Ocean and from the east coast out into the Atlantic Ocean, the most useful satellite imagery will now be provided by the two (2) NOAA satellites currently in use. The next GOES satellite to be sent up to replace GOES East will be sometime in 1985 which will return the satellite to normal.

Basically, two types of imagery are available and when combined give a great deal of information about clouds. Through interpretation, the analyst can determine the type of cloud, the temperature of cloud tops (from this the approximate height of the cloud) and the thickness of cloud layers. From this information, the analyst gets a good idea of the associated weather.

One type of imagery is visible imagery (see figure 3-4 and 3-6). With a visible picture we are looking at clouds and the earth reflecting sunlight to the satellite sensor. The greater the reflected sunlight reach-

FIGURE 3-5. GOES Infra-red Imagery.

FIGURE 3-6. NOAA Visible Imagery.

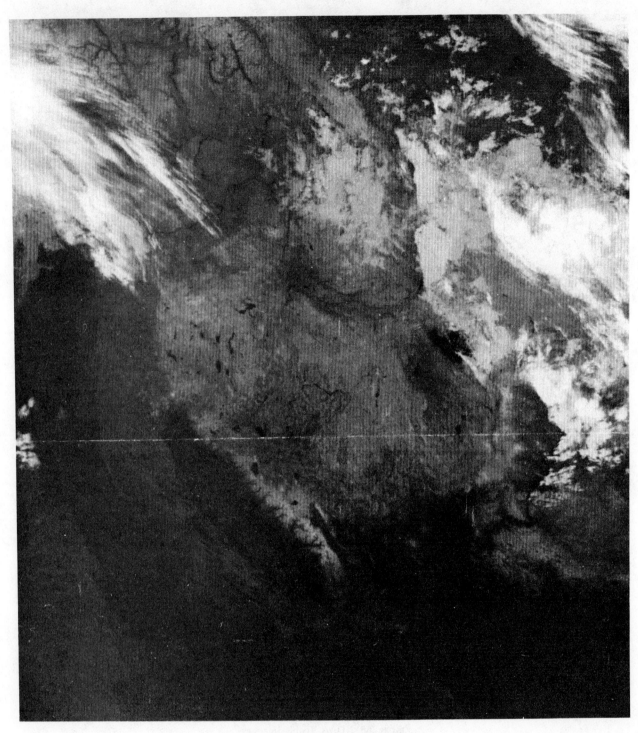

FIGURE 3-7. NOAA Infra-Imagery.

ing the sensor, the whiter the object is on the picture. The amount of reflectivity reaching the sensor depends upon the height, thickness and ability of the object to reflect sunlight. We can generalize that clouds are much more reflective than the earth and so clouds will show up white on the picture, especially thick clouds. Thus, the visible picture is primarily used to determine 1) the presence of clouds and 2) the type of cloud from shape and texture. There are no visible pictures at night from the GOES satellite. The NOAA satellite does take night visible pictures through the ability of the moon to reflect sunlight to the earth at night.

The second type of imagery is infra-red imagery (IR) (see figure 3-5 and 3-7). With an IR picture we are looking at heat radiation being emitted by the clouds and earth. The images show temperature differences between cloud tops and the ground, as well as, temperature gradations of cloud tops and temperature gradations along the earth's surface. Ordinarily, cold temperatures are displayed as light gray or white, making high clouds appear whitest, but various enhancements are sometimes used to sharply illustrate important temperature changes. IR pictures are used to determine cloud top temperatures and thus the approximate height of the cloud. IR pictures are available both day and night. From this, one can see the importance of using visible and IR imagery together when interpreting clouds.

Operationally, at WSFOs and FSSs, pictures are received once every 30 minutes. In these they can see development and dissipation of weather, such as fog and convection, all over the country. Much of this is not visible from reporting points.

Section 4
AVIATION WEATHER FORECASTS

Good flight planning considers forecast weather. This section explains the following aviation forecasts:

1. TERMINAL FORECASTS
   (a). Domestic (FT)
   (b). International (ICAO TAF)
2. AREA FORECAST (FA)
3. TWEB ROUTE FORECAST AND SYNOPSIS
4. CONVECTIVE SIGMET (WST)
5. SIGMET AND AIRMET (WS and WA)
6. WINDS AND TEMPERATURES ALOFT FORECAST (FD)
7. SPECIAL FLIGHT FORECASTS
8. CENTER WEATHER SERVICE UNIT (CSWU) PRODUCTS

Also discussed are the following general forecasts which may aid in flight planning:

1. HURRICANE ADVISORY (WH)
2. CONVECTIVE OUTLOOK (AC)
3. SEVERE WEATHER WATCH BULLETIN (WW)
4. ALERT SEVERE WEATHER WATCH MESSAGE (AWW)

TERMINAL FORECASTS (FT AND TAF)

FT (Domestic)

A terminal forecast (FT) is a description of the surface weather expected to occur at an airport. The forecast cloud heights and amounts, visibility, weather and wind relate to flight operations within 5 nautical miles of the center of the runway complex. The term vicinity (VCNTY) covers the area from 5 miles beyond the center of the runway complex out to 25 miles. For example, "TRW VCNTY" means thunderstorms are expected to occur between 5 and 25 miles from the station. Scheduled forecasts are issued by WSFOs for their respective areas (figures 1-4 and 1-4A) 3 times daily and are valid for 24 hours. Issue and valid times are according to time zones (see Section 14). The format of the FT is essentially the same as that of the SA report.

Example of a Terminal Forecast (FT)

STL 251010 C5 X 1/2S-BS 3325G35 OCNL C0 X 0S+BS. 16Z C30 BKN 3BS 3320 CHC SW-. 22Z 30 SCT 3315. 00Z CLR. 04Z VFR WND..

To aid in the discussion, we have divided the forecast into the following elements lettered "a" through "i"

STL   251010   C5 X   1/2   S-BS   3325G35   OCNL C0 X 0S+BS.
 a.   b.   c.   d.   e.   f.   g.

16Z C30 BKN 3BS 3320 CHC SW-. 22Z 20SCT 3315. 00Z CLR.
                             h.

04Z VFR WND..
      i.

a. Station identifier. "STL" identifies St. Louis, MO. The forecast is for St. Louis.
b. Date-time group. "251010" is date and valid times. The forecast is valid beginning on the 25th day of the month at 1000Z valid until 1000Z the following day.
c. Sky and ceiling. "C5 X" means ceiling 500 feet, sky obscured. The letter "C" always identifies a forecast ceiling layer. Cloud heights are always referenced to ground level. Sky cover designators and height coding are identical to those used in the SA reports.

The summation of cloud layers concept is used here as in the SA. Total sky cover at each level from a ground observers point of view is determined instead of the individual cloud layers at each level. Thus, there will not be a "SCT" layer above a "BKN" layer because total sky cover can only increase with each succeeding higher level.

d. Visibility. "1/2" means visibility 1/2 mile. Visibility is in statute miles and fractions. Absence of a visibility entry specifically implies visibility more than 6 statute miles.
e. Weather and obstructions to vision. "S-BS" means light snow and blowing snow. These elements are in symbols identical to those used in SA reports and are entered only when forecast.
f. Wind. "3325G35" means wind from 330 degrees at 25 knots gusting to 35 knots which is the same as in the SAs. Omission of a wind entry specifically implies wind less than 6 knots.
g. Remarks. "OCNL C0 X 0S+BS" means occa-

4-1

sional conditions of ceiling zero, sky obscured, visibility zero, heavy snow and blowing snow. Remarks may be added to more completely describe forecast weather by indicating variations from prevailing conditions. See table 4-10 for the definitions of the variability terms used in remarks. In this case the occasional conditions described above are expected to occur with a greater than 50% probability but for less than 1/2 of the time period from 1000Z to 1600Z. Thus, c. through f. in the above example are the prevailing conditions from 1000Z to 1600Z with g. specifying variations from the prevailing conditions during the same time period. Also, "LLWS" will be included in the remarks section if low level wind shear is forecast.

In the remarks section, caution must be used in interpreting the data when two (2) or more variable conditions are forecast for different time periods within the same forecast group. If a phenomenon in the remarks is forecast for only a portion of the time period of the forecast group, the time period for the phenomenon is indicated immediately after the phenomenon. Example:

00Z C30 OVC OCNL 12 OVC 02Z-04Z CHC TRW−. 06Z etc.

This forecast says prevailing conditions will be ceiling 3000 feet from 00Z to 06Z. A ceiling of 1200 feet is expected between 02Z and 04Z for less than 1/2 of the 2 hours. The chance of thunderstorms with light rainshowers is for the period 00Z to 06Z.

    h. Expected changes. When changes are expected, preceding conditions are followed by a period before the time and conditions of the expected change. " 16Z C30 BKN 3BS 3320 CHC SW−. 22Z 30 SCT 3315. 00Z CLR." means by 1600 Z the prevailing conditions will be 3000 Broken, visibility 3 miles, blowing snow, wind 330 degrees at 20 knots and a 30 to 50% chance of light snow showers. By 2200Z, the prevailing conditions will change to 3,000 scattered, visibility more than 6 miles (implied) and wind 330 degrees at 15 knots. By 0000Z, prevailing conditions will become sky clear, visibility more than 6 miles and wind less than 10 knots (implied).

    i. 6-hour categorical outlook. The last 6 hours of the forecast is a categorical outlook. "04Z VFR WND.." means that from 0400Z until 1000Z (the end of the forecast period) weather will be ceiling more than 3,000 feet or none and visibility greater than 5 miles (VFR) with wind 25 knots or stronger. The double period (..) signifies the end of the forecast for the specific terminal.

See table 4-1 for a list of applicable categories and remarks that may be appended to the category "VFR" to better describe the expected conditions. The cause(s) of below VFR categorical outlooks must be stated. Below VFR categories can be due to ceilings only (CIG), restrictions to visibility only (TRW F etc.). See table 4-2 for examples. The term "WND" is not included if winds (sustained or gusts) are forecast to be less than 25 knots.

TABLE 4-1. Categories

Category	Definition
LIFR	Low IFR-ceiling less than 500 feet and/or visibility less than 1 mile.
IFR	Ceiling 500 to less than 1,000 feet and/or visibility 1 to less than 3 miles.
MVFR	Marginal VRF-ceiling 1,000 to 3,000 feet and/or visibility 3 to 5 miles inclusive.
VFR	No ceiling or ceiling greater than 3,000 feet and visibility greater than 5 miles.

Remarks that may be appended to VFR at forecaster's discretion	Definition
VFR CIG ABV 100	Ceiling greater than 10,000 feet and visibility greater than 5 miles.
VFR NO CIG	Cloud coverage less than 6/10 or thin clouds and visibility greater than 5 miles.
VFR CLR	Cloud coverage less than 1/10 and visibility greater than 5 miles.

TABLE 4-2. Examples of categorical groupings

Example	Definition
LIFR CIG	Low IFR due to low ceiling *only*.
IFR F	IFR due to visibility restricted by fog *only*
MVFR CIG H K	Marginal VFR due *both* to low ceiling and to visibility restricted by haze and smoke.
IFR CIG R WND	IFR due *both* to low ceiling and to visibility restricted by rain, wind expected to be 25 knots or greater.

Scheduled FT Collectives

The heading of an FT collective identifies the message as an FT along with a 6-digit date-time group giving the transmission time. For example, "FT130940 means a collective transmitted on the 13th at 0940Z. A collective FT message will usually be broken down into states, ie "TX 130940" would be followed by a group of FTs for terminals in the state of Texas.

Out of Sequence FTs.

A delayed, corrected, or amended FT is identified in the message rather than in the heading. The following are a delayed FT for Binghamton NY, a corrected FT for Memphis TN, and an amended FT for Lufkin TX.

BGM FT RTD 131615 1620Z 100 SCT 250 SCT 1810. 18Z 50 SCT 100 SCT 1913 CHC C30 BKN 3TRW AFT 20Z. 03Z 100 SCT C250 BKN. 09Z VFR..

MEM FT COR 132222 2230Z 40 SCT 300 SCT CHC TRW. 02Z CLR. 16Z VFR..

LFK FT AMD 1 131410 1425Z C8 OVC 4F OVC OCNL BKN. 15Z 20 SCT 250-BKN. 19Z 40 SCT 120 SCT CHC C30 BKN 3TRW. 04Z MVFR CIG F..

Note in each forecast a time group following the valid period which is the issue time. For example, the BGM delayed forecast was issued at 1620Z and not at the scheduled issuance time of 1440Z. This changes the beginning of the valid forecast period from 1500Z to 1600Z.

An FT will not be issued unless two consecutive observations are received from the station. So a routine delayed FT (RTD) is usually due to a station not being on a 24 hour observing schedule with the first two observations of the day being received after the regularly scheduled FT issuance time. If the time is known when observations usually end for the day the phrase "NO AMDTS AFT (TIME) Z" will be appended to the *last* scheduled FT. For example, let's say Oklahoma City's last observation for the day is 03Z. To indicate that no amendments will be available after 03Z due to lack of observations, the last regularly scheduled FT might look like this:

OKC 172222 CLR 16Z VFR CLR. NO AMDTS AFT 03Z..

A new FT will not be issued until two (2) consecutive observations are received.

A corrected FT is necessary due to a typographical error in the FT.

An amended FT is necessary for a situation in which the forecast has to be revised due to significant changes in the weather. Note also that the amended forecast for LFK has the entry "AMD 1". Amended FTs for each terminal are numbered sequentially starting after each scheduled forecast.

International Civil Aviation Organization (ICAO) Terminal Forecast (TAF)

Terminal forecasts for long overwater international flights (TAF) are in an alphanumeric code. They are scheduled four times daily for 24-hour periods beginning at 0000Z, 0600Z, 1200Z and 1800Z (see Section 14 for issuance times).

Format. The TAF is a series of groups made up of digits and letters. An individual group is identified by its position in the sequence, by its alphanumeric coding or by a numerical indicator. Listed below are a few contractions used in the TAF. Some of the contractions are followed by time entries indicated by "tt" or "tttt" or by probability, "pp".

Significant weather change indicators

GRADU tttt—A gradual change occurring during a period in excess of one-half hour. "tttt" are the beginning and ending times of the expected change to the nearest hour; i.e., "GRADU 1213" means the transition will occur between 1200Z and 1300Z.

RAPID tt —A rapid change occurring in one-half hour or less. "tt" is the time to the nearest hour of the change; i.e., "RAPID 23" means the change will occur about 2300Z.

Variability terms—indicate that short time period variations from prevailing conditions are expected with the total occurrence of these variations less than 1/2 of the time period during which they are called for.

TEMPO tttt—Temporary changes from prevailing conditions of less than one hour duration in each instance. There may be more than one (1) instance for a specified time period. "tttt" are the earliest and latest times during which the temporary changes are expected; i.e., "TEMPO 0107" means the temporary changes may occur between 0100Z and 0700Z.

INTER tttt —Changes from prevailing conditions are expected to occur frequently and briefly. "tttt" are the earliest and latest times the brief changes are expected; i.e., "IN-

TER 1518" means that the brief, but frequent, changes may occur between 1500Z and 1800Z. INTER has shorter and more frequent changes than TEMPO.

Probability

PROB pp —Probability of conditions occurring. "pp" is the probability in percent; i.e., "PROB 20" means a 10 or 20% probability of the conditions occurring. "PROB 40" means a 30 to 50% inclusive probability.

Cloud and weather terms

CAVOK —No clouds below 50,000 feet or below the highest minimum sector altitude whichever is greater, and no cumulonibus. Visibility *6 miles* or greater. No precipitation, thunderstorms, shallow fog or low drifting snow.

WX NIL —No significant weather (no precipitation, thunderstorms or obstructions to vision).

SKC —Sky clear.

Following is a St. Louis MO forecast in TAF code. It is the same as the preceding FT example on page 4-1 except that it begins 2 hours later.

KSTL 1212 33025/35 0800 71SN 9//05 INTER 1215 0000 75XXSN 9//000 GRADU 1516 33020 4800 38BLSN 7SC030 PROB 40 85SNSH GRADU 2122 33015 9999 WX NIL 3SC030 RAPID 00 VRB05 9999 SKC GRADU 0304 24015/25 CAVOK ⓪

The forecast is broken down into the elements lettered "a" to "l" to aid in the discussion. Not included in the example but explained at the end are three optional forecast groups for "m" icing, "n" turbulence and "o" temperature.

KSTL 1212 33025/35 0800 71SN 9//005
 a.   b.   c.   d.   e.   f.

INTER 1215 0000 75XXSN 9//000
g.

GRADU 1516 33020 4800 38BLSN 7SC030
h.

PROB 40 85SNSH
i.

GRADU 2122 33015 9999 WX NIL 3SC030
j.

RAPID 00 VRB05 9999 SKC
k.

GRADU 0304 24015/25 CAVOK ⓪
l.

a. Station identifier. The TAF code uses ICAO 4-letter station identifiers. In the contiguous 48 states the 3-letter identifier is prefixed with a "K"; i.e., the 3-letter identifier for Seattle is SEA while the ICAO identifier is KSEA. Elsewhere, the first two letters of the ICAO identifier tell what region the station is in. "MB" means Panama/Canal Zone (MBHO is Howard AFB); "MI" means Virgin Islands (MISX is St. Croix); "MJ" is Puerto Rico (MJSJ is San Juan); "PA" is Alaska (PACD is Cold Bay); " PH" is Hawaii (PHTO is Hilo).

b. Valid time. Valid time of the forecast follows station identifier. "1212" means a 24-hour forecast valid from 1200Z until 1200Z the following day.

c. Wind. Wind is forecast usually by a 5-digit group giving degrees in 3 digits and speed in 2 digits. When wind is expected to be 100 knots or more, the group is 6-digits with speed given is 3 digits. When speed is gusty or variable, peak speed is separated from average speed with a slash. For example, in the KSTL TAF, "33025/35" means wind 330 degrees, average speed 25 knots, peak speed 35 knots. A group "160115/130" means wind 160 degrees, 115 knots, peak speed 130 knots. "00000" means calm; "VRB" followed by speed indicates direction variable; i.e., "VRB10" means wind direction variable at 10 knots.

d. Visibility. Visibility is in meters. TABLE 4-5 is a table for converting meters to miles and fractions. "0800" means 800 meters converted from table to 1/2 mile.

e. Significant weather. Significant weather is decoded using TABLE 4-3. Groups in the table are numbered sequentially. Each number is followed by an acronym suggestive of the weather; you can soon learn to read most of the acronyms without reference to the table. Examples: "177TS", thunderstorm; "18SQ", squall; "31SA", sandstorm; "60RA", rain; "85SNSH", snow shower. "XX" freezing rain. In the KSTL forecast, "71SN" means light snow. The TAF encodes only the single most significant type of weather; the U.S. domestic FT permits encoding of multiple weather types. See TABLE 4-4 to convert weather from FT to TAF.

f. Clouds. A cloud group is a 6-character group. The first digit is coverage in octas (eighths) of the individual cloud layer only. The summation of cloud layer to determine total sky cover from a ground observers point of view is NOT used. See TABLE 4-6. The two letters identify cloud type as shown in the same table. The last

TABLE 4-3. TAF weather codes

Code	Simple Definition	Detailed Definition
04FU	Smoke	Visibility reduced by smoke. No visibility restriction.*
05HZ	Dust haze	Visibility reduced by haze. No visibility restriction.
06HZ	Dust haze	Visibility reduced by dust suspended in the air but wind not strong enough to be adding more dust. No visibility restriction.*
07SA	Duststorm, sandstorm, rising dust or sand	Visibility reduced by dust suspended in the air and wind strong enough to be adding more dust. No well developed dust devils, duststorm or sandstorm. Visibility 6 miles or less.
08PO	Dust devil	Basically the same as 07SA but with well developed dust devils. Visibility 6 miles or less.
10BR	Mist	Fog, ground fog or ice fog with visibility 5/8 to 6 miles.
11MIFG	Shallow fog	Patchy shallow fog (less than 6 feet deep and coverage less than half) with visibility in the fog less than 5/8 mile.
12MIFG	Shallow fog	Shallow fog (less than 6 feet deep with more or less continuous coverage) with visibility in the fog less than 5/8 miles.
17TS	Thunderstorm	Thunderstorm at the station but with no precipitation.
18SQ	Squall	No precipitation. A sudden increase of at least 15 knots in average wind speed, sustained at 20 knots or more for at least one (1) minute. Not an easy thing to forecast!
19FC	Funnel cloud	Used to forecast a tornado, funnel cloud or waterspout at or near the station. Also not easy to forecast and likely to be overshadowed by some other more violent weather such as thunderstorms.
30SA	Duststorm, sandstorm, rising dust	Duststorm or sandstorm, visibility 5/16 to less than 5/8 mile, decreasing in intensity.
31SA		Basically the same as 30SA but with no change in intensity.

* While this may seem to be contradictory, it means that while visibility *is* restricted, the *amount* of the restriction is not limited.

TABLE 4-3. TAF weather codes (Cont.)

Code	Simple Definition	Detailed Definition
32SA	or sand	Basically the same as 30SA but increasing in intensity.
33XXSA	Heavy duststorm	Severe duststorm or sandstorm, visibility less than 5/16 mile, decreasing in intensity.
34XXSA	or sandstorm	Basically the same as 33XXSA but with no change in intensity.
35XXSA		Basically the same as 33XXSA but increasing in intensity.
36DRSN	Low drifting	Low drifting snow (less than 6 feet) with visibility in drifting snow less than 5/16 miles.
37DRSN	snow	Low drifting snow (less than 6 feet) with visibility in drifting snow less than 5/16 miles.
38BLSN	Blowing snow	Blowing snow (more than 6 feet) with visibility 5/16 to 6 miles.
39BLSN		Blowing snow (more than 6 feet deep) with visibility 5/16 to 6 miles.
40BCFG	Fog patches	Distant fog (not at station).
41BCFG		Patchy fog at the station, visibility in the fog patches less than 5/8 miles.
42FG		Fog at the station, visibility less than 5/8 mile, sky visible, fog thinning.
43FG		Fog at the station, visibility less than 5/8 mile, sky not visible, fog thinning.
44FG	Fog	Fog at the station, visibility less than 5/8 mile, sky visible, no change in intensity.
45FG		Fog at the station, visibility less than 5/8 mile, sky not visible, no change in intensity.
46FG		Fog at the station, visibility less than 5/8 mile, sky visible, fog thickening.
47FG		Fog at the station, visibility less than 5/8 mile, sky not visible, fog thickening.

NOTE: In code figures 40 through 47, "fog" includes both fog and ice fog.
See FMH No.1 (Surface Observations) for definitions of precipitation intensities.

TABLE 4-3. TAF weather codes (Cont.)

Code	Simple Definition	Detailed Definition
48FZFG	Freezing	Fog depositing rime ice, visibility less than 5/8 mile, sky visible.
49FZFG	fog	Fog depositing rime ice, visibility less than 5/8 mile, sky not visible.
50DZ		Light intermittent drizzle.
51DZ		Light continuous drizzle.
52DZ	Drizzle	Moderate intermittent drizzle.
53DZ		Moderate continuous drizzle.
54XXDZ	Heavy	Heavy intermittent drizzle.
55XXDZ	drizzle	Heavy continuous drizzle.
56XXDZ	Freezing drizzle	Light freezing drizzle.
57XXFZDZ	Heavy freezing drizzle	Moderate or heavy freezing drizzle.
58RA		Mixed rain and drizzle, light.
59RA		Mixed rain and drizzle, moderate or heavy.
60RA	Rain	Light intermittent rain.
61RA		Light continuous rain.
62RA		Moderate intermittent rain.
63RA		Moderate continuous rain.
64XXRA	Heavy	Heavy intermittent rain.
65XXRA	rain	Heavy continuous rain.
66FZRA	Freezing rain	Freezing rain or mixed freezing rain and freezing drizzle, light.
67XXFZRA	Heavy freezing rain	Freezing rain or mixed freezing rain and freezing drizzle, moderate or heavy.
68RASN	Rain and snow	Mixed rain and snow or drizzle and snow, light.
69XXRASN	Heavy rain and snow	Mixed rain and snow or drizzle and snow, moderate or heavy.
70SN		Light intermittent snow.
71SN	Snow	Light continuous snow.
72SN		Moderate intermittent snow.
73SN		Moderate continuous snow.
74XXSN	Heavy	Heavy intermittent snow.
75XXSN	snow	Heavy continuous snow.

NOTE: See FMH No.1 (Surface Observations) for definitions of precipitation intensities.

TABLE 4-3. TAF weather codes (Cont.)

Code	Simple Definition	Detailed Definition
77SG	Snow grains	Snow grains, any intensity. May be accompanied by fog or ice fog.
79PE	Ice pellets	Ice pellets, any intensity. May be mixed with some other precipitation.
80RASH	Showers	Light rain showers.
81XXSH	Heavy	Moderate or heavy rain showers.
82XXSH	showers	Violent rain showers (more than 1 inch per hour or 0.1 inch in 6 minutes).
83RASN	Showers of rain and snow	Mixed rain showers and snow showers. Intensity of both showers is light.
84XXRASN	Heavy showers of rain and snow	Mixed rain showers and snow showers. Intensity of either shower is moderate or heavy.
85SNSH	Snow showers	Light snow showers.
86XXSNSH	Heavy snow showers	Moderate or heavy snow showers.
87GR		Light ice pellet showers. There may also be rain or mixed rain or snow.
88GR	Soft hail	Moderate or heavy ice pellet showers. There may also be rain or mixed rain and snow.
89GR	Hail	Hail, not associated with a thunderstorm. There may be also rain or mixed rain and snow.
90XXGR	Heavy hail	Moderate or heavy hail, not associated with a thunderstorm. There may also be rain or mixed rain or snow.
91RA	Rain	Light rain or light rain shower at the time of the forecast *and* thunderstorms during the preceding hour but not at the time of the forecast.
92XXRA	Heavy rain	Basically the same as 91RA but the intensity of the rain or rain shower is moderate or heavy.
93GR	Hail	Basically the same as 91RA but the precipitation is light snow or snow showers, or light mixed rain and snow or rain showers and snow showers, or light ice pellets or ice pellet showers.

TABLE 4-3. TAF weather codes (Cont.)

Code	Simple Definition	Detailed Definition
94XXGR	Heavy hail	Basically the same as 93GR but the intensity of any precipitation is moderate or heavy.
95TS	Thunderstorm	Thunderstorm with rain or snow, or a mixture of rain and snow, but no hail, ice pellets or snow pellets.
96TSGR	Thunderstorm with hail	Thunderstorm with hail, ice pellets or snow pellets. There may also be rain or snow, or mixed rain and snow.
97XXTS	Heavy thunderstorm	Severe thunderstorm with rain or snow, or a mixture of rain and snow, but no hail, ice pellets or snow pellets.
98TSSA	Thunderstorm with duststorm or sandstorm	Thunderstorm with duststorm or sandstorm. There may also be some form of precipitation with the thunderstorm.
99XXTSGR	Heavy thunderstorm with hail	Basically the same as 97XXTS but in addition to everything else there is hail.

three digits are cloud height in hundreds of feet above ground level (AGL). In the KSTL TAF, "9//005" means sky obscured (9), clouds not observed (//), vertical visibility 500 feet (005). The TAF may include as many cloud groups as necessary to describe expected sky condition.

g. **Expected variation from prevailing conditions.** Variations from prevailing conditions are identified by the contractions INTER and TEMPO as defined earlier. In the KSTL TAF, "INTER 1215 0000 75XXSN 9//000" means intermittently from 1200Z to 1500Z (1215) visibility zero meters (0000) or zero miles, heavy snow (75XXSN), sky obscured, clouds not observed, vertical visibility zero (9//000).

h, i, j, k, and l. An expected change in prevailing conditions is indicated by the contraction GRADU and RAPID as defined earlier. In the KSTL TAF, "GRADU 1516 33020 4800 38BLSN 7SC030" means a gradual change between 1500Z and 1600Z to wind 330 degrees at 20 knots, visibility 4,800 meters or 3 miles (TABLE 4-5), blowing snow, 7/8 stratocumulus (TABLE 4-6) at 3000 feet AGL. "PROB 40 85SNSH" means there is a 30 to 50% probability that light snow showers will occur between 1600Z and 2100Z. "GRADU 2122 33015 9999 WX NIL 3SC030" means a gradual change between 2100Z and 2200Z to wind 330 degrees at 15 knots, visibility 10 kilometers or more (more than 6 miles), no significant weather, 3/8 stratocumulus at 3000 feet. "RAPID 00 VRB05 9999 SKC" means a rapid change about 0000Z to wind direction variable at 5 knots, visibility more than 6 miles, sky clear. "GRADU 0304 24015/25 CAVOK ⓞ" means a gradual change between 0300Z and 0400Z to wind 240 degrees at 15 knots, peak gust to 25 knots with CAVOK conditions. ⓞ means end of message.

m. **Icing.** An icing group may be included. It is a 6-digit group. The first digit is 6 identifying it as an icing group. The second digit is the type of ice accretion from TABLE 4-7 top. The next three digits are height of the base of the icing layer in hundreds of feet (AGL). The last digit is the thickness of the layer in *thousands* of feet. For example, let's decode the group "680304". "6" indicates an icing forecast; "8" indicates severe icing in cloud; "030" says the base of the icing is at 3,000 feet (AGL); and "4" specifies a layer 4,000 feet thick.

n. **Turbulence.** A turbulence group also may be included. It also is a 6-digit group coded the same as the icing group except a "5" identifies the group as a turbulence forecast. Type of turbulence is from TABLE 4-7 bottom. For example, decoding the group "590359", "5" identifies a turbulence forecast; "9" specifies frequent severe turbulence in cloud (TABLE 4-7); "035" says the base of the turbulent layer is 3,500 feet (AGL); "9" specifies that the turbulence layer is 9,000 feet thick.

When either an icing layer or a turbulent layer is expected to be more than 9,000 feet thick, multiple groups are used. The top specified in one group is coincident with the base in the following group. Let's assume the forecaster expects frequent turbulence from the surface to 45,000 feet with the most hazardous turbulence at mid-levels. This could be encoded "530005 550509 592309 553209 554104". While you most likely will never see such a complex coding with this many groups, the flexible TAF code permits it.

o. **Temperature.** A temperature code is seldom included in a terminal forecast. However, it may be included if critical to aviation. It may be used to alert the pilot to high density altitude or possible frost when on the ground. The temperature group is identified by the digit "0". The next two (2) digits are the time to the nearest whole hour (GMT) to which the fore-

TABLE 4-4. Converting significant weather from U.S. terms to WMO terms.

Express the TAF code equivalents as shown in the appropriate column below

When forecasting any of these in U.S. domestic code		Precipitation & Intensity		
		Light	Moderate	Heavy
a	89GR			
BD or BN (vsby 5/16 to 1/2 mi)	31SA			
BD or BN (vsby 0 to 1/4 mi)	34XXSA			
BS (vsby 6 mi or less)	38BLSN			
D (vsby 6 mi or less)	06HZ			
GF (vsby 1/2 mi or less)	44FG			
H (vsby 6 mi or less)	05HZ			
F or IF (vsby 1/2 mi or less)	45FG			
F, GF or IF (vsby 5/8 to 6 mi)	10BR			
IP	79PE			
IPW	87GR			
K	04FU			
L		51DZ	53DZ	55DZ
R		61RA	63RA	64RA
RS		68RASN	68RASN	69XXRASN
RW		80RASH	80RASH	81XXSH
RWSW		83RASN	83RASN	84XXRASN
S		71SN	73SN	75XXSN
SG	77SG			
SP	87GR			
SW		85SNSH	85SNSH	86XXSNSH
ZL		56FZDZ	56FZDZ	57XXFZDZ
ZR		66FZRA	67XXFZRA	67XXFZRA
TRW− or TRW	95TS			
TRW+	95TS			INTER 81XXSH
				INTER 82XXSH*
TRW−A or TRWA	96TSGR			
T+RW	97XXTS			
T+RW+	97XXTS			INTER 81XXSH
	97XXTS			INTER 82XXSH*
T+RWA	99XXTSGR			
T+RW+A	99XXTSGR			INTER 81XXSH
	99XXTSGR			INTER 82XXSH*

* INTER 82XXSH is to be encoded in a TAF only when a violent rainshower (at least 1 inch of rain per hour or 0.10 inch in 6 minutes) is forecast.
NOTE: Conversions from TAF to FT will not be exact in some cases due to a lack of a one to one relationship.

cast temperature applies. The last two (2) digits are temperature in degrees Celsius. A minus temperature is preceded by the letter "M". Examples: "02137" means temperature at 2100Z is expected to be 37 degrees Celsius (about 99 degrees F); "012M02" means temperature at 1200Z is expected to be minus 2 degrees Celsius. A forecast may include more than one temperature group.

Example of a DOMESTIC FA

DFWH FA 041040
HAZARDS VALID UNTIL 042300
OK TX AR LA TN MS AL AND CSTL WTRS

TABLE 4-5. Visibility conversion—TAF code to miles

Meters	Miles	Meters	Miles	Meters	Miles
0000	0	1200	3/4	3000	1 7/8
0100	1/16	1400	7/8	3200	2
0200	1/8	1600	1	3600	2 1/4
0300	3/16	1800	1 1/8	4000	2 1/2
0400	1/4	2000	1 1/4	4800	3
0500	5/16	2200	1 3/8	6000	4
0600	3/8	2400	1 1/2	8000	5
0800	1/2	2600	1 5/8	9000	6
1000	5/8	2800	1 3/4	9999	>6

TABLE 4-6. TAF cloud code

Code	Cloud amount		Cloud type
0	0 (clear)	CI	Cirrus
1	1 octa or less but not zero	CC	Cirrocumulus
2	2 octas	CS	Cirrostratus
3	3 octas	AC	Altocumulus
4	4 octas	AS	Altostratus
5	5 octas	NS	Nimbostratus
6	6 octas	SC	Stratocumulus
7	7 octas or more but not 8 octas	ST	Stratus
		CU	Cumulus
8	8 octas (overcast)	CB	Cumulonimbus
9	Sky obscured or cloud amount not estimated	//	Cloud not visible due to darkness or obscuring phenomena

TABLE 4-7. TAF icing and turbulence

Figure code	Amount of ice accretion (TAF group 6)
0	No icing
1	Light icing
2	Light icing in cloud
3	Light icing in precipitation
4	Moderate icing
5	Moderate icing in cloud
6	Moderate icing in precipitation
7	Severe icing
8	Severe icing in cloud
9	Severe icing in precipitation

Figure code	Turbulence (TAF group 5)
0	None
1	Light Turbulence
2	Moderate turbulence in clear air, infrequent
3	Moderate turbulence in clear air, frequent
4	Moderate turbulence in cloud, infrequent
5	Moderate turbulence in cloud, frequent
6	Severe turbulence in clear air, infrequent
7	Severe turbulence in clear air, frequent
8	Severe turbulence in cloud, infrequent
9	Severe turbulence in cloud, frequent

FLT PRCTNS...TURBC...TN AL AND CSTL
WTRS   ...ICG...TN
         ...IFR...TX
TSTMS IMPLY PSBL SVR OR GTR TURBC
SVR ICG AND LLWS NON MSL HGTS NOTED
BY AGL OR CIG
THIS FA ISSUANCE INCORPORATES THE
FOLLOWING AIRMETS STILL IN EFFECT
...NONE.

DFWS FA 041040
SYNOPSIS VALID UNTIL 050500
AT 11Z RDG OF HI PRES ERN TX NWWD TO
CNTRL CO WITH HI CNTR OVR ERN TX. BY
05Z HI CNTR MOVS TO CNTRL LA.

DFWI FA 041040
ICING AND FRZLVL VALID UNTIL 042300
TN
FROM SLK TO HAT TO MEM TO ORD TO SLK
OCNL MDT RIME ICGIC ABV FRZLVL TO
100. CONDS ENDING BY 17Z. FRZLVL 80

CHA SGF LINE SLPG TO 120 S OF A IAH MAF
LINE.

DFWT FA 041040
TURBC VALID UNTIL 042300
TN AL AND CSTL WTRS
FROM SLK TO FLO TO 90S MOB TO MEI TO
BUF TO SLK
OCNL MDT TURBC 250-380 DUE TO JTSTR.
CONDS MOVG SLOLY EWD AND CONTG
BYD 23Z.

DFWC FA 041040
SGFNT CLOUD AND WX VALID UNTIL 042300
...OTLK 042300-050500
IFR...TX
FROM SAT TO PSX TO BRO TO MOV TO SAT
VSBY BLO 3F TIL 15Z.
OK AR TX LA MS AL AND CSTL WTRS
80 SCT TO CLR EXCP VSBY BLO 3F TIL 15Z
OVR PTNS S CNTRL TX. OTLK...VFR.
TN
CIGS 30-50 BKN 100 VSBYS OCNLY 3-5F
BCMG AGL 40-50 SCT TO CLR BY 19Z.
OTLK...VFR.

DOMESTIC AREA FORECAST (FA)

An area forecast (FA) is a forecast of general weather conditions over an area the size of several states. It is used to determine forecast enroute weather and to interpolate conditions at airports which do not have FTs issued. Figure 1-5, section 1, maps the FA areas. FAs are issued 3 times a day by the National Aviation Weather Advisory Unit (NAWAU) in Kansas City for each of the 6 areas in the contiguous 48 states. In Alaska, FAs are issued by the WSFOs in Anchorage, Fairbanks, and Juneau for their respective areas (figure 1-5A). The WSFO in Honolulu issues FAs for Hawaii (figure 1-5A). See Section 14 for issuance times.

Each FA consists of a 12 hour forecast plus a 6 hour outlook. All times are Greenwich Mean Time (GMT). All distances except visibility are in nautical miles. Visibility is in statute miles.

The FA is comprised of 5 sections, HAZARDS/FLIGHT PRECAUTIONS (H), SYNOPSIS (S), ICING (I), TURBULENCE (T) (AND LOW LEVEL WIND SHEAR, if applicable), and SIGNIFICANT CLOUDS AND WEATHER (C). Each section has an unique communications header which allows replacement of individual sections, due to amendments or corrections, instead of replacing the entire FA. For example (using the FA example on the previous page):

DFWH FA 041040

states that this section of the FA which deals with the hazards section (H) has been issued on the 4th day of the month at 1040Z for the Dallas-Fort Worth (DFW) forecast area.

HAZARDS/FLIGHT PRECAUTIONS (H) Section

A 12 hour forecast that identifies and locates aviation weather hazards which meet Inflight Advisory criteria and thunderstorms that are forecast to be at least scattered in area coverage. These hazards include IFR conditions, icing (ICG), turbulence (TURBC), mountain obscurations (MTN OBSCN), and thunderstorms (TSTMS). A discussion of the hazards section from the above DFW FA continues:

HAZARDS VALID UNTIL 042300

This states that the hazards listed may be valid for the 12 hour forecast time period 11Z to 23Z or for only a portion of the time period. If a specific time period for a hazard is to be stated, it will be stated in the appropriate subsequent section.

OK TX AR LA TN MS AL AND CSTL WTRS

This identifies the states and geographical area that make up the DFW forecast area. This statement will be found in all DFW FAs and does *not* outline the hazard areas.

FLT PRCTNS...TURBC...TN AL AND CSTL WTRS...ICG...TN ...IFR...TX

This states that TURBC, ICG, and IFR conditions are forecast within the 12 hour period for the listed states, within the designated FA boundary. The forecasts are stated in subsequent sections. If no hazards are expected, "NONE EXPECTED" will be written.

TSTMS IMPLY PSBL SVR OR GTR TURBC SVR ICG AND LLWS

is found in all FAs as a reminder of the hazards existing in all thunderstorms. Thus, these thunderstorm associated hazards are not spelled out within the body of the FA.

NON MSL HGTS NOTED BY AGL OR CIG

You will find that this statement is contained in all FAs to alert the user that heights, for the most part, are *above sea level*. All heights are in hundreds of feet. For example, 30 BKN 100 HIR TRRN OBSCD means bases of broken clouds 3,000 feet with tops 10,000 feet MSL. Terrain above 3,000 will be obscured. The tops of clouds and icing/freezing level heights are *always* MSL.

Heights *above ground level* will be denoted in either of two ways:
(1) Ceilings by definition are above ground. Therefore, the contraction "CIG" indicates above ground. For example, "CIGS GENLY BLO 10" means that ceilings are expected to be generally below 1,000 feet.
(2) The contraction "AGL" means above ground level. Therefore, "AGL" 20 SCT" means scattered clouds with bases 2,000 feet above ground level.

Thus, if the contraction "AGL" or "CIG" is not denoted, height is automatically above MSL.

THIS FA ISSUANCE INCORPORATES THE FOLLOWING AIRMETS STILL IN EFFECT ...NONE.

A statement contained in all FAs stating that any AIRMET in effect at the time of FA issuance is incorporated into the FA. The AIRMET is then cancelled. In this example, no AIRMETs were in effect when this FA was issued.

SYNOPSIS Section

A brief summary of the location and movement of fronts, pressure systems, and circulation patterns for an 18 hour period.

ICING Section

A forecast of non-thunderstorm related icing of light or greater intensity for up to 12 hours. If a trace or less of icing is expected, the remark "NO SGFNT ICING EXPCD" is used. Otherwise, the location of each icing phenomenon is specified in a separate paragraph containing (1) The affected states or areas within the designated FA boundary, (2) The VOR points outlining the *entire* area of icing and (3) the type, intensity, and heights of the icing. For example:

TN
FROM SLK TO HAT TO MEM TO ORD TO SLK
OCNL MDT RIME ICGIC ABV FRZLVL TO 100

identifies Tennessee as the only state within the DFW forecast area that is forecast to experience OCNL MDT RIME ICGIC ABV FRZLVL TO 100. The entire area of icing is enclosed by the VOR points (see figure 4-2 on page 4-38) listed and includes states covered in other FA areas. The lowest freezing level heights are specified in a separate statement in hundreds of feet MSL.

TURBULENCE/LOW LEVEL WIND SHEAR Section

This section forecasts non-thunderstorm related turbulence of moderate or greater intensity and low level wind shear for up to 12 hours. If moderate or greater turbulence is not expected, the remark "NO SGFNT TURBC EXPCD" is used. Otherwise, each location of turbulence phenonmenon is specified in a separate paragraph using the same format found in the icing section. For example:

TN AL AND CSTL WTRS
FROM SLK TO FLO TO 90S MOB TO MEI TO BUF TO SLK
OCNL MDT TURBC 250-380 DUE TO JTSTR. CONDS MOVG SLOLY EWD AND CONTG BYD 23Z.

Low level wind shear (LLWS) potential, when forecast, is included as a separate paragraph. It is omitted from the DFW FA example above since none was forecast. If LLWS had been forecast, an example of how it would appear follows:

LLWS POTENTIAL OVR WRN NY FROM 03Z-05Z DUE TO STG WMFNT.

Note: If the total area to be affected by any hazard (ICG, TURBC, MTN OBSCN, IFR, TSTMS) during the forecast period (as outlined by VOR's) is very large, it could be only a portion of this total area may be affected at any one time.

As a specific example, let us examine the above forecast of jetstream turbulence from 250-380. Figure 4-1 outlines the total areas to be affected during the 12 hour period. The forecast for CONDITIONS MOVING SLOWLY EASTWARD AND CONTINUING BEYOND 23Z tells us that the phenomenon will move slowly out of the western portion of the outlined area and into the eastern portion of the outlined area during the 12 hour period. By late in the time period, the western portion will be free of significant turbulence but the eastern portion will continue to experience significant turbulence beyond 23Z.

SIGNIFICANT CLOUD AND WEATHER Section

A 12 hour forecast, in broad terms, of clouds and weather significant to flight operations plus a 6 hour categorical outlook. Table 4-8 defines the contractions and compares them to the designators used in the FT. Surface visibility and obstructions to vision are included when forecast visibility is 5 miles or less. Precipitation, thunderstorms, and sustained winds of 30 knots or greater are always included when forecast. Table 4-9 gives expected coverage indicated by the terms "isolated", "widely scattered", "scattered", and "numerous". Table 4-10 identifies variability terms used.

TABLE 4-8. Contractions in FAs

Contraction	FT Designator	Definition
CLR	CLR	Sky clear
SCT	SCT	Scattered
BKN	BKN	Broken
OVC	OVC	Overcast
OBSCD	X	Obscured
PTLY OBSCD	−X	Partly obscured
	−	Thin
CIG	C	Ceiling

TABLE 4-9. Area coverage of showers and thunderstorms

Adjective	Coverage
Isolated	Single cells (no percentage)
Widely scattered	Less than 25% of area affected
Scattered	25 to 54% of area affected
Numerous	55% or more of area affected

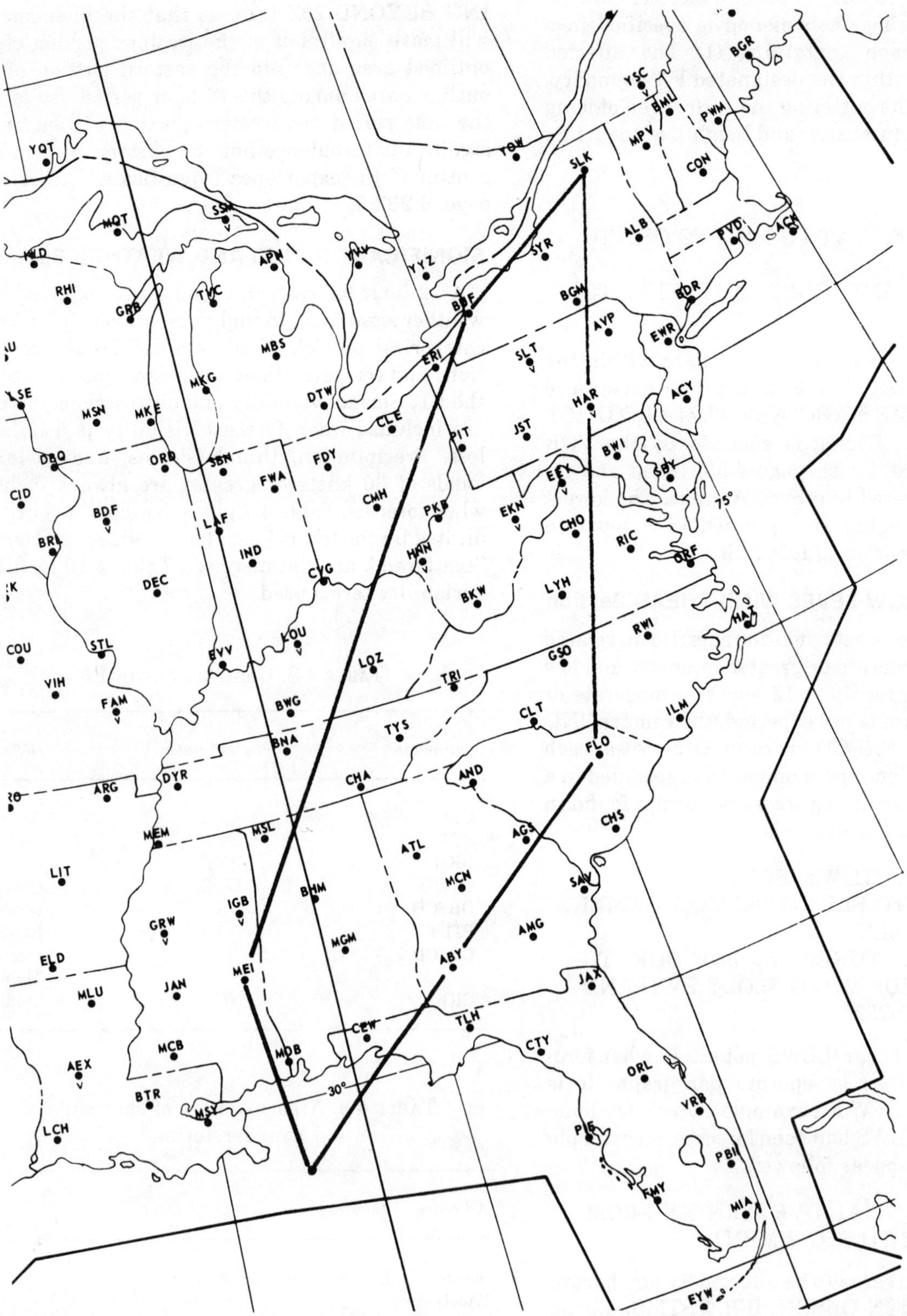

FIGURE 4-1. Area of jetstream turbulence in FA example.

TABLE 4-10. Variability terms

Term	Description
OCNL	Greater than 50% probability of the phenomenon occurring but for less than 1/2 of the forecast period
CHC	30 to 50% probability (precipitation only)
SLGT CHC	10 or 20% probability (precipitation only)

The SGFNT CLOUD AND WX section is usually several paragraphs. The breakdown may be by states, by well known geographical areas, or in reference to location and movement of a pressure system or front. Figure 4-2 is a map to assist you in identifying geographical areas. An example would be

OK AR TX LA MS AL AND CSTL WTRS
80 SCT TO CLR EXCP VSBY BLO 3F TIL
15Z OVR PTNS S CNTRL TX.

A categorical outlook, identified by "OTLK", is included for each area breakdown. For example,

OTLK...VFR.

Categorical outlooks of IFR and MVFR can be due to ceilings only (CIG), restrictions to visibility only (TRW F etc.), or a combination of both (CIG TRW F etc.). For example, "OTLK...VFR BCMG MVFR CIG F AFT 09Z" means the weather is expected to be VFR becoming MVFR (marginal VFR) due to low ceilings and visibilities restricted by fog after 0900Z. "WIND" is included in the outlook if winds, sustained or gusty, are expected to be 30 knots or greater. For definitions of each category, refer to the section on the weather depiction chart.

FLT PRCTNS from the hazards section which are described in this section, as IFR and MTN OBSCN, will be specified in a separate paragraph at the beginning of the section using the same format as the icing and turbulence sections. TSTMS, however, will *not* be included as a separate paragraph. For example,

IFR...TX
FROM SAT TO PSX TO BRO TO MOV TO
SAT
VSBY BLO 3F TIL 15Z.

AMENDED AREA FORECASTS

Amendments to the FA are issued as needed. Only that section of the FA being revised is transmitted as an amendment. Area forecasts are also amended and updated by inflight advisories but the FA section affected will always be kept current by amending it. An amended FA is identified by "AMD", a corrected FA is identified by "COR", and a delayed FA is identified by "RTD".

TWEB ROUTE FORECASTS AND SYNOPSIS

The TWEB Route Forecast is similar to the Area Forecast (FA) except information is contained in a route format. Forecast sky cover (height and amount of cloud bases), cloud tops, visibility (including vertical visibility), weather, and obstructions to vision are described for a corridor 25 miles either side of the route. Cloud bases and tops are always MSL unless noted. Ceilings are always above ground.

The Synopsis is a brief statement of frontal and pressure systems affecting the route during the forecast valid period.

The TWEB Route Forecasts are prepared by the WSFOs for more than 300 selected short-leg and cross-country routes over the contiguous U.S. (figure 1-6 and figure 1-7, section 1). WSFOs prepare synopses for the routes in their areas. These forecasts go into the Transcribed Weather Broadcasts (TWEB) and the Pilot's Automatic Telephone Weather Answering Service (PATWAS) transcriptions described in section 1. Individual route forecasts and synopses are also available by request/reply teletypewriter through any FSS or WSO.

The TWEB Route Forecasts and Synopses are issued by the WSFOs three times per day according to time zone. See Section 14 for issuance times. The early morning and midday forecasts are valid for 12 hours and the evening forecast for 18 hours. This schedule provides 24-hour coverage with most frequent updating during the hours of greatest general aviation activity.

Example of a TWEB Synopsis:

BIS SYNS 252317. LO PRES TROF MVG ACRS ND TDA AND TNGT. HI PRES MVG SEWD FM CANADA INTO NWRN ND BY TNGT AND OVR MST OF ND BY WED MRNG.

BIS—Bismarck ND. WSFO issuing Synopsis and Route Forecasts
SYNS—Synopsis for the area covered by the Route Forecasts
25—25th day of the month
2317—Valid 23Z on the 25th to 17Z on the 26th (18 hours)

(Rest of message)

—LOW PRESSURE TROUGH MOVING ACROSS NORTH DAKOTA TODAY AND TONIGHT. HIGH PRESSURE MOVING SOUTHEASTWARD FROM CANADA INTO NORTHWESTERN NORTH DAKOTA BY TONIGHT AND OVER MOST OF NORTH DAKOTA BY WEDNESDAY MORNING.

Example of a TWEB Route Forecast:

249 TWEB 252317 GFK-MOT-ISN. GFK VCNTY CIGS AOA 5 THSD TILL 12Z OTRW OVR RTE CIGS 1 TO 3 THSDS VSBY 3 TO 5 MI IN LGT SNW WITH CONDS BRFLY LWR IN HVYR SNW SHWRS

    249—Route number
TWEB—TWEB Route Forecast
    25—25th day of month
    2317—Valid 23Z on the 25th to 17Z on the 26th (18 hours)
GFK-MOT-ISN-Route: Grand Forks to Minot to Williston ND
(Rest of message)
—GRAND FORKS VICINITY CEILINGS AT OR ABOVE 5000 FEET UNTIL 1200Z OTHERWISE OVER ROUTE CEILINGS 1 TO 3 THOUSAND FEET VISIBILITY 3 TO 5 MILES IN LIGHT SNOW WITH CONDITIONS BRIEFLY LOWER IN HEAVIER SNOW SHOWERS.

When visibility is not stated it is implied to be greater than 6 miles.

Because of their varied accessibility and route format, these forecasts are important and useful weather information available to the pilot for flight operations and planning. You should become familiar with them and use them regularly.

INFLIGHT ADVISORIES (WST, WS, WA)

Inflight advisories are unscheduled forecasts to advise enroute aircraft of development of potentially hazardous weather. All inflight advisories in the conterminous U.S. (48 states) are issued by the National Aviation Weather Advisory Unit (NAWAU) in Kansas City. In Alaska, the three WSFOs (Anchorage, Fairbanks, and Juneau) issue inflight advisories for their respective areas. The WSFO in Honolulu issues advisories for Hawaii. All heights are referenced to MSL, except in the case of ceilings (CIG) which indicates above ground level. The advisories are of three types—CONVECTIVE SIGMET (WST), SIGMET (WS), and AIRMET (WA). All inflight advisories use the same location identifiers (either VORs or well known geographic areas) to describe the hazardous weather areas (see figures 4-2 and 4-3).

CONVECTIVE SIGMET (WST)

CONVECTIVE SIGMETs are issued in the conterminous U.S. for any of the following:
1. Severe thunderstorm due to a) surface winds greater than or equal to 50 knots or b) hail at the surface greater than or equal to 3/4 inches in diameter or c) tornadoes.
2. Embedded thunderstorms.
3. Line of thunderstorms.
4. Thunderstorms greater than or equal to VIP level 4 affecting 40% or more of an area at least 3000 square miles.

Any CONVECTIVE SIGMET implies severe or greater turbulence, severe icing, and low level wind shear. A CONVECTIVE SIGMET may be issued for any convective situation which the forecaster feels is hazardous to all categories of aircraft.

CONVECTIVE SIGMET bulletins are issued for the Eastern (E), Central (C) and Western (W) United States. The areas separate at 87 and 107 degrees west longitude with sufficient overlap to cover most cases when the phenomenon crosses the boundaries. Thus, a bulletin will usually be issued only for the area where the bulk of observed and forecast conditions are located. Bulletins are issued hourly at H+55. Special bulletins are issued at any time as required and updated at H+55. If no criteria meeting a CONVECTIVE SIGMET are observed or forecast, the message "CONVECTIVE SIGMET...NONE" will be issued for each area at H+55. Individual SIGMETs for each area are numbered sequentially (01-99) each day, beginning at 00Z. A continuing CONVECTIVE SIGMET phenomenon will be reissued every hour at H+55 with a new number. The text of the bulletin consists of either an observation and a forecast or just a forecast. The forecast is valid for up to 2 hours.

EXAMPLE

The following are examples of CONVECTIVE SIGMET bulletins for the Central U.S., For the Western U.S., they would be numbered 17W, 18W, and 19W while for the Eastern U.S., they would be numbered 17E, 18E, and 19E.

MKCC WST 221655
CONVECTIVE SIGMET 17C
KS OK TX
VCNTY GLD-CDS LINE
NO SGFNT TSTMS RPRTD
FCST TO 1855Z
LINE TSTMS DVLPG BY 1755Z WILL MOV EWD 30-35 KTS THRU 1855Z
HAIL TO 1 1/2 IN PSBL.

MKCC WST 221655
CONVECTIVE SIGMET 18C
SD NE IA
FROM FSD TO DSM TO GRI TO BFF TO FSD
AREA TSTMS WITH FEW EMBDD CELLS MOVG FROM 2725 TOPS 300
FCST TO 1855Z
DSPTG AREA WILL MOV EWD 25 KNTS.

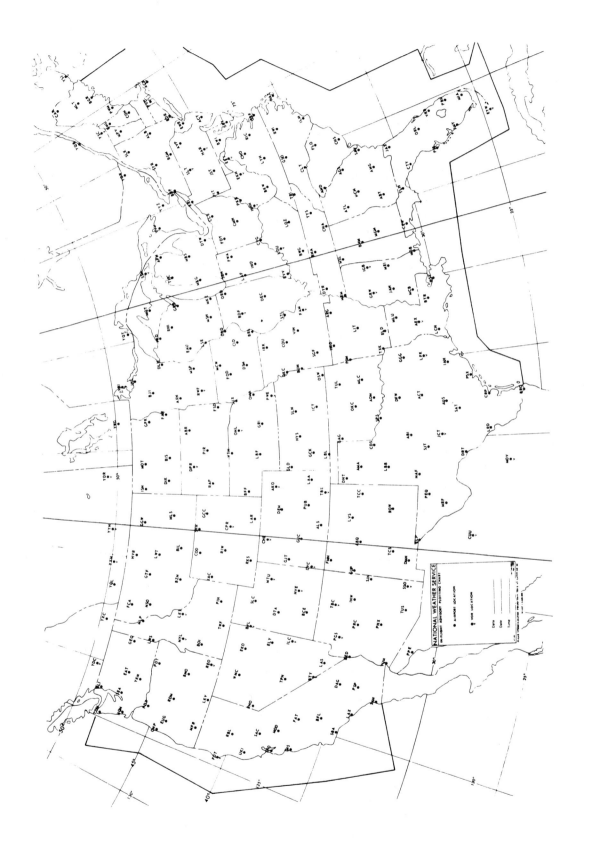

FIGURE 4-2. Inflight Weather Advisory Location Identifier (VORs)

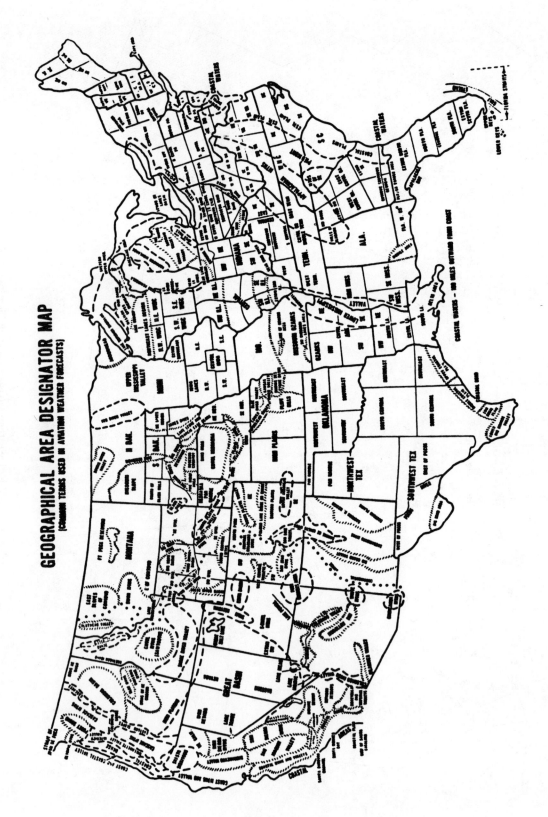

FIGURE 4-3. Geographical areas and terrain features. Forecasts often best locate weather by reference to terrain.

```
MKCC WST 221755
CONVECTIVE SIGMET 19C
KS OK
FROM 30E TO 20E GAG
DVLPG LINE TSTMS 25 MI WIDE MOVG
FROM 2315 TOPS 450.
HAIL TO 1 IN...WIND GUSTS TO 55.
FCST TO 1955Z
LINE WILL CONT INTSFYG AND MOV
NEWD 25-30 KTS THRU 1955Z. HAIL TO 2 IN
PSBL.
```

The first example is a bulletin issued at 1655Z on the 22nd day of the month. It is the 17th CONVECTIVE SIGMET of the day in the Central U.S.. Although no significant thunderstorm activity is noted at 1655Z, a line of thunderstorms is expected to develop by 1755Z near a Goodland-Childress line (see figure 4-2 for complete list of VORs) in the states of Kansas, Oklahoma, and Texas and move eastward 30-35 knots possibly producing 1 1/2 inch hail thru 1855Z.

Note that 19C is an update and re-issuance of 17C.

SIGMET (WS)/AIRMET (WA)

A SIGMET advises of weather potentially hazardous to all aircraft other than convective activity. In the conterminous U.S., items covered are:
1. Severe icing
2. Severe or extreme turbulence
3. Duststorms, sandstorms, or volcanic ash lowering visibilities to less than three (3) miles

In Alaska and Hawaii there are no CONVECTIVE SIGMETs. In these states we add:
4. Tornadoes
5. Lines of thunderstorms
6. Embedded thunderstorms
7. Hail greater than or equal to 3/4 inch diameter

An **AIRMET** is for weather that may be hazardous to single engine, other light aircraft, and VFR pilots. AIRMETs should be read by all pilots. The items covered are:
1. Moderate icing
2. Moderate turbulence
3. Sustained winds of 30 knots or more at the surface
4. Ceilings less than 1000 feet and/or visibility less than 3 miles affecting over 50% of the area at one time.
5. Extensive mountain obscurement

These SIGMET/AIRMET items are considered "widespread" because they must be affecting or be forecast to affect an area of at least 3000 square miles at any one time. However, if the total area to be affected during the forecast period (as outlined by VOR's) is very large, it could be that only a small portion of this total area would be affected at any one time. An example would be a 3000 square mile phenomenon forecast to move across an area totaling 25,000 square miles during the forecast period. For a specific example, see turbulence section in the Area Forecast (FA).

SIGMETs/AIRMETs are issued for 6 areas corresponding to the FA areas (figure 1-5) with a maximim forecast period of 4 hours for SIGMETs and 6 hours for AIRMETs. If conditions persist beyond the forecast period, the SIGMET/AIRMET must be updated and reissued.

A phenomenon is identified by an alphabetic designator in the series ALFA through NOVEMBER for SIGMETs and OSCAR through ZULU for AIRMETs. Issuances for the same phenomenon will be sequentially numbered. For example, ALFA 1 is the first issuance for a SIGMET phenomenon, ALFA 2 is the second issuance for the same phenomenon, etc. The first issuance of a SIGMET will be labeled UWS (Urgent Weather SIGMET). UWS will also be used in subsequent issuances at the discretion of the forecaster. For an AIRMET the first issuance will be OSCAR 1, etc. All designators in the series will be used before starting over again with ALFA and OSCAR. The alphabetic designator assigned to a phenomenon will be retained until the phenomenon ends, even when moving from one area into another. For example, the first issuance in the CHI area for phenomenon moving in from the SLC area will be SIGMET ALFA 3 if the previous two issuances, ALFA 1 and ALFA 2 had been in the SLC area. Since no two different phenomena across the country can have the same alphabetic designator at the same time, all 6 areas must use the same ALFA through NOVEMBER and OSCAR through ZULU series.

EXAMPLES:

```
DFWA UWS 051710
SIGMET ALFA 1 VALID UNTIL 052110
AR LA MS
FROM MEM TO 30N MEI TO BTR TO MLU TO
MEM
OCNL SVR ICING ABV FRZLVL EXPCD.
FRZLVL 080 E TO 120 W.
CONDS CONTG BYD 2100Z.

SFOB WS 100130
SIGMET BRAVO 2 VALID UNTIL 100530
OR WA
FROM SEA TO PDT TO EUG TO SEA
OCNL MOGR CAT BTN 280 AND 350 EXPCD
DUE TO JTSTR. CONDS BGNG AFT 0200Z
CONTG BYD 0530Z AND SPRDG OVR CNTRL
ID BY 0400Z.
```

```
MIAP WAS 151900
AIRMET PAPA 2 VALID UNTIL 160100
GA FL
FROM SAV TO JAX TO CTY TO TLH TO SAV
MDT TURBC BLO 100 EXPCD. CONDS IPVG
AFT 160000Z.
```

The first example above is a SIGMET bulletin issued for the DFW area at 1710Z on the 5th and is valid until 2110Z (Note maximum forecast period of 4 hours for a SIGMET). The designator ALFA identifies the phenomenon, in this case, severe icing. This is the first issuance of the SIGMET as indicated by "UWS" and "ALFA 1". The affected states *within* the DFW area are Arkansas, Louisiana, and Mississippi. VORs (see figure 4-2) outline the *entire* area to be affected (irrespective of FA boundaries) by severe icing during the forecast period. Freezing level data and notation that conditions are expected to continue beyond 4 hours are included. See TABLE 4-10 for defintions of variability terms.

It is important to note that AIRMETs will only be issued for conditions meeting AIRMET criteria that are *not* already forecast in the area forecast (FA). SIGMETs will be issued whether or not they are forecast in the FA.

WINDS AND TEMPERATURES ALOFT FORECAST (FD)

Winds and temperatures aloft are forecast for specific locations in the contiguous U.S. as shown in figure 1-3, section 1. FD forecasts are also prepared for a network of locations in Alaska as shown in figure 1-3-A, section 1. Forecasts are made twice daily based on 00Z and 12Z data for use during specific time intervals.

Below is a sample FD message containing a heading and six FD locations. The heading always includes the time during which the FD may be used (1700-2100Z in the example) and a notation "TEMPS NEG ABV 24000". Since temperatures above 24,000 feet are always negative, the minus sign is omitted.

```
FD KWBC 151640
BASED ON 151200Z DATA
VALID 151800Z FOR USE 1700-2100Z TEMPS NEG ABV 24000

FT     3000   6000    9000    12000   18000   24000   30000   34000   39000
ALA           2420    2635-08 2535-18 2444-30 245945  246755  246862
AMA    2714   2725+00 2625-04 2531-15 2542-27 265842  256352  256762
DEN           2321-04 2532-08 2434-19 2441-31 235347  236056  236262
HLC    1707-01 2113-03 2219-07 2330-17 2435-30 244145  244854  245561
MKC    0507  2006+03 2215-01 2322-06 2338-17 2348-29 236143  237252  238160
STL    2113  2325+07 2332+02 2339-04 2356-16 2373-27 239440  730649  731960
```

FORECAST LEVELS

The line labelled "FT" shows 9 of 11 standard FD levels. The 45,000 and 53,000 foot levels are not transmitted on teletypewriter circuits but are available in the communications system. The pilot may request these levels from the FSS briefer or NWS meteorologist. Through 12,000 feet the levels are true altitude, 18,000 feet and above are pressure altitude. The FD locations are transmitted in alphabetical order.

Note that some lower level groups are omitted. No winds are forecast within 1,500 feet of station elevation. No temperatures are forecast for the 3,000 foot level or for a level within 2,500 feet of station elevation.

DECODING

A 4-digit group shows wind direction (reference true north) and windspeed. Look at the St. Louis (STL) forecast for 3,000 feet. The group 2113 means wind from 210 degrees at 13 knots. The first two digits give direction in tens of degrees and the second two speed in knots.

A 6-digit group includes forecast temperature. In the STL forecast, the coded group for 9,000 feet is 2332+02 which is wind from 230 degrees at 32 knots and temperature +2 degrees Celsius.

Encoded windspeed 100 to 99 knots have 50 added to the direction code and 100 subtracted from the speed. The STL forecast for 39,000 feet is "731960". Wind is from 230 degrees at 119 knots and temperature −60 degrees Celsius.

How do you recognize when coded direction has been increased by 50? Coded direction (in tens of degrees) range from 01 (010 degrees) to 36 (360 degrees). Thus, a coded direction of more than "36" indicates winds 100 knots or more. Coded direction with speeds of over 100 knots range from 51 through 86.

If windspeed is forecast at 200 knots or greater, the wind group is coded as 199 knots; i.e., "7799" is decoded 270 degrees at 199 knots or greater.

When the forecast speed is less than 5 knots, the coded group is "9900" and read, "LIGHT AND VARIABLE".

Examples of decoding FD winds and temperatures:

Coded	Decoded
9900+00	Wind light and variable, temperature 0 degree Celsius
2707	270 degrees at 7 knots
850552	350 degrees (85-50=35) at 105 knots (05+100=105), temperature −52 degrees Celsius

SPECIAL FLIGHT FORECAST

When planning a special category flight and scheduled forecasts are insufficient to meet your needs, you may request a special flight forecast through any FSS or WSO. Special category flights are hospital or rescue flights; experimental, photographic or test flights; record attempts; and mass flights such as air tours, air races and fly-aways from special events.

Make your request far enough in advance to allow ample time for preparing and transmitting the forecast. Advance notice of 6 hours is desirable. In making a request, give the:

1. Aircraft mission
2. Number and type of aircraft
3. Point of departure
4. Route of flight (including intermediate stops, destination, alternates)
5. Estimated time of departure
6. Time enroute
7. Flight restrictions (such as VFR, below certain altitudes, etc.)
8. Time forecast is needed

The forecast is written in plain language contractions as in the examples:

SPL FLT FCST ABQ-PHOTO MISSION-ABQ 121500Z. THIN CI CLDS AVGG LESS THAN TWO TENTHS CVR. VSBY MORE THAN 30. WNDS AND TEMPS ALF AT FLT ALTITUDE 2320+03. ABQ WSFO 121300Z.

SPL FLT OTLK MKC-RST 062100Z–062400Z. CIG 2 THSD OVC OR BTR. WNDS ALF AT FLT ALTITUDE 2320. MKC WSFO 052300Z.

CENTER WEATHER SERVICE UNIT (CWSU) PRODUCTS

Center Weather Service Unit products are issued by the CWSU meteorologist located in the ARTCCs. Coordination between the CWSU meteorologist and the nearby NWS WSFO is extremely important because both will address the same event. If time permits, coordination should take place before the CWSU meteorologist issues a product.

METEOROLOGICAL IMPACT STATEMENT (MIS)

A Meteorological Impact Statement (MIS) is an unscheduled traffic/flight operations planning forecast of conditions expected to begin generally 4 to 12 hours after issuance. This enables the impact of expected weather conditions to be included in traffic control related decisions of the near future.

A MIS will be issued when the following three (3) conditions are met:

1. if any one of the following conditions are forecast.
   a. convective SIGMET criteria
   b. moderate or greater icing and/or turbulence
   c. heavy or freezing precipitation
   d. low IFR conditions
   e. surface winds/gusts 30 knots or greater
   f. low level wind shear within 2000 feet of the surface
   g. volcanic ash, dust, or sandstorm
2. if impact occurs on air traffic flow within the ARTCC area of responsibility.
3. if forecast lead time (the time between issuance and outset of a phenomenon), in the forecaster's judgement, is sufficient to make issuance of a Center Weather Advisory (CWA) unnecessary.

An example of a MIS:

ZKC MIS 02 031800Z-040100Z
ENROUTE...NONE.
TERMINAL...STL...SCT OCNL BKN 015-020 BKN OCNL OVC 040-080 CHC LVL 3/4 TSTMS. WIDELY SCT TSTMS MOVG E 20 KTS. 22Z MVFR CONDS WITH BKN-OVC CIGS 008-015 CHC VSBYS 4-6 FOG/HAZE. SFC WINDS 250-280 10-15 KTS SHFTG 280-310 AFT 22Z.

This MIS from Kansas City MO ARTCC is the 2nd issuance of the day; issued at 1800Z on the 3rd and is valid until 0100Z on the 4th.

CENTER WEATHER ADVISORY (CWA)

A Center Weather Advisory (CWA) is an unscheduled inflight flow control, air traffic and air crew *advisory* for use in anticipating and avoiding adverse weather conditions in the enroute and terminal areas. The CWA is *not* a flight planning forecast but a *nowcast* for conditions beginning within the next two (2) hours. Maximum valid time of a CWA is two (2) hours, i.e. no more than 2 hours between issuance time and "valid until time". If conditions are expected to continue beyond the valid period, a statement will be included in the advisory.

A CWA may be issued for the following three (3) situations:

1. as a supplement to an *existing* inflight advisory or area forecast (FA) section for the purpose of improving or updating the definition of the phenomenon in terms of location, movement, extent, or intensity *relevant* to the ARTCC area of responsibility. This is important for the following reason. A SIGMET for severe turbulence issued by NAWAU may outline the entire ARTCC area for the total four (4) hour valid

period but may only be covering a relatively small portion of the ARTCC area at any one time during the four (4) hour period.

2. when an inflight advisory has not yet been issued but conditions meet inflight advisory criteria based on current pilot reports and the information must be disseminated sooner than NAWAU can issue the inflight advisory. In this case of an impending SIGMET, the CWA will be issued as urgent "UCWA" to allow the fastest possible dissemination.

3. when inflight advisory criteria is not met but conditions are or will shortly be adversely affecting the safe flow of air traffic within the ARTCC area of responsibility.

Format of a CWA heading:
ARTCC Designator and Phenomenon number (numbers 1 through 6 used for replaceability) /"CWA" /issuance number (2 digits) /inflight advisory alphanumeric designator (if applicable) /date and time issued /"—" /valid until time.

Examples of a CWA:

ZFW3 CWA 03 032140-2340
ISOLD SVR TSTM OVR MLU MOVG SWWD 10 KTS. TOP 610. WND GUSTS TO 55 KTS. HAIL TO 1 INCH RPRTD AT MLU. SVR TSTM CONTG BYND 2340.

ZKC1 CWA 01/ALFA 4 121528-1728
NUMEROUS RPRTS OF MDT TO SVR ICG 080-090 30 MILE RADIUS OF STL. LGT OR NEG ICG RPRTD 040-120 RMNDR OF ZKC AREA.

HURRICANE ADVISORY (WH)

When a hurricane threatens a coast line, but is located at least 300 NM off shore, an abbreviated hurricane advisory (WH) is issued to alert aviation interests. The advisory gives location of the storm center, its expected movement, and maximum winds in and near the storm center. It does not contain details of associated weather. Specific ceilings, visibilities, weather, and hazards are found in the area and terminal forecasts and inflight advisories.

An example of an abbreviated aviation hurricane advisory:

MIA WH 181010
HURCN IONE AT 1000Z CNTRD 29.4N 74.2W OR 400 NMI E OF JACKSONVILLE FL EXPCTD TO MOV N ABT 12 KT. MAX WNDS 110 KT OVR SML AREA NEAR CNTR AND HURCN WNDS WITHIN 55-75 NM.

CONVECTIVE OUTLOOK (AC)

A convective outlook (AC) describes the prospects for general thunderstorm activity during the following 24 hours. Areas in which there is a high, moderate or slight risk of severe thunderstorms are included as well as areas where thunderstorms may approach severe limits (approaching means winds greater than or equal to 35 knots but less than 50 knots and/or hail greater than or equal to 1/2 inch in diameter but less than 3/4 inch). Refer to the "Severe Weather Outlook Chart" for "risk" definitions. Forecast reasoning is also included in all ACs.

Outlooks are transmitted by the National Severe Storm Forecast Center (NSSFC) in Kansas City MO at 0800Z and 1500Z, and between February 1 and August 31 at 1930Z. Forecasts in each AC are valid until 1200Z the next day and are used for preparing and updating the Severe Weather Outlook Chart.

Use the outlook primarily for planning flights later in the day.

Severe thunderstorm criteria:
a. wind greater than or equal to 50 knots at the surface or
b. hail greater than or equal to 3/4 inch diameter at the surface or
c. tornadoes

The following is a convective outlook:

MKC AC 031500
VALID 031500—041200Z

THERE IS A MDT RISK OF SVR TSTMS THIS AFTN AND EVE PTNS ERN AL..ERN TN..ERN KY..WV..PA..NY ..VT..NH..MA..CT..NJ..DE..MD.. VA..NC..SC..GA. AREA IS TO RT OF LN FM DHN MGM HSV LOZ HTS PIT SYR MPV PSM BOS GON ACY SBY RDU AGS ABY DHN.

GEN TSTM ACTVY TO RT OF LN FM BPT MLU MEM OWB TOL..CONTD JAX CTY. ALSO TO RT OF LN FM CDC ELY BYI IDA LND RWL DEN CEZ CDC.
UPR LVL LOW NR MLI WITH TROF EXTNDG SWD INTO ERN TX EXPCD TO MOV NEWD. VRY STG UPR LVL JET FM GGG DAY PWM EXPCD TO CONT MOVG SLOLY EWD PROVIDING UPR LVL SHEAR AND DVRG FIELDS. NARROW BAND OF INSTBLTY RANGES FM MINUS 7 IN AL TO MINUS 4 IN SERN NY. RPDLY MOVG CELLS EXPCD TO MOV THRU WARM AND MOIST AIR ORIENTED FM AL TO WV TO ERN NY. ISOLD TRW PSBL THIS AFTN FM SRN ID INTO CO AS WK UPR TROF MOV ACRS AREA OF MARGINAL INSTBLY.

SEVERE WEATHER WATCH BULLETIN (WW)

A severe weather watch bulletin (WW) defines areas of possible severe thunderstorms or tornado activity. The bulletins are issued by the National Severe Storm Forecast Center at Kansas City MO. WWs are unscheduled and are issued as required.

A severe thunderstorm watch describes expected severe thunderstorms and a tornado watch states that the additional threat of tornadoes exists in the designated watch area.

In order to alert the WSFOs, WSOs, CWSUs, and FSS's, and other users, a preliminary message called the Alert Severe Weather Watch message (AWW) is sent before the main bulletin.

Example of a preliminary message:

MKC AWW 161755
WW 279 SEVERE TSTM NY PA NJ
161830Z—17000Z AXIS..70 STATUTE MILES EITHER SIDE OF LINE..
10W MSS.20E ABE
HAIL SURFACE AND ALOFT..2 INCHES.
WIND GUSTS..65 KNOTS.
MAX TOPS TO 540. MEAN WIND VECTOR 19020. REPLACES WW 278.. OH PA NY.

The Severe Weather Watch Bulletin format:

A. Type of severe weather watch, watch area, valid time period, type of severe weather possible, watch axis, meaning of a watch, and a statement that persons be on the lookout for severe weather.
B. Other watch information..references to previous watches.
C. Phenomena, intensities, hail size, wind speeds (knots), maximum CB tops, and estimated cell movement (mean wind vector).
D. Cause of severe weather.
E. Information on updating ACs.

Example of a Severe Weather Watch Bulletin (WW)

MKC WW 161800
BULLETIN IMMEDIATE BROADCAST REQUESTED
SEVERE THUNDERSTORM WATCH NUMBER 279
NATIONAL WEATHER SERVICE KANSAS CITY MO
200 PM EDT THU JUN 16 1983

A..THE NATIONAL SEVERE STORMS FORECAST CENTER HAS ISSUED A SEVERE THUNDERSTORM WATCH FOR

EASTERN HALF OF NEW YORK
NORTHEASTERN PENNSYLVANIA
NORTHERN NEW JERSEY

FROM 230 PM EDT UNTIL 800 PM EDT THIS THURSDAY AFTERNOON AND EVENING

LARGE HAIL AND DAMAGING THUNDERSTORM WINDS ARE POSSIBLE. IN THESE AREAS. THE SEVERE THUNDERSTORM WATCH AREA IS ALONG AND 70 STATUTE MILES EITHER SIDE OF A LINE FROM 10 MILES WEST OF MASSENA NEW YORK TO 20 MILES EAST OF ALLENTOWN PENNSYLVANIA

REMEMBER....A SEVERE THUNDERSTORM WATCH MEANS CONDITIONS ARE FAVORABLE FOR SEVERE THUNDERSTORMS IN AND CLOSE TO THE WATCH AREA.
PERSONS IN THESE AREAS SHOULD BE ON THE LOOKOUT FOR THREATENING WEATHER CONDITIONS AND LISTEN FOR LATER STATEMENTS AND POSSIBLE WARNINGS.

B..OTHER WATCH INFORMATION..THIS SEVERE THUNDERSTORM WATCH REPLACES SEVERE THUNDERSTORM WATCH NUMBER 278. WATCH NUMBER 278 WILL NOT BE IN EFFECT AFTER 230 PM EDT.

C..A FEW SVR TSTMS WITH HAIL SFC AND ALF TO 2 IN. EXTRM TURBC AND SFC WIND GUSTS TO 65 KT. A FEW CBS WITH MAX TOPS TO 540. MEAN WIND VECTOR 19020.

D..TSTMS EXPCD TO INCRS RPDLY IN ZONE OF WK SFC CONVG WHERE AMS HAS LI OF MINUS 8.

E..OTR TSTMS...WW MAY BE RQD SOON FOR PTNS ERN WY NERN CO AND WRN NEB. UPDATE AC TO INCL GEN TSTM ACTVY IN SRN FL THIS AFTN TO RT OF LINE FROM FMY PBI.

Status reports are issued as needed to show progress of storms and to delineate areas no longer under the threat of severe storm activity. Cancellation bulletins are issued when it becomes evident that no severe weather will develop or that storms have subsided and are no longer severe.

When tornadoes or severe thunderstorms have developed, local WSOs and WSFOs issue local warnings.

Section 5
SURFACE ANALYSIS CHART (observations)

A surface analysis is commonly referred to as a surface weather chart. In the contiguous 48 states a computer prepared chart covering these states and adjacent areas is transmitted every three hours. Areas with facsimile receive surface weather charts at regularly scheduled intervals. Figure 5-1 is a section of a surface weather chart and Figure 5-2 illustrates the symbols depicting fronts and pressure centers. The following explains the contents of the chart.

VALID TIME

Valid time of the chart corresponds to the time of the plotted observations. A date-time group in Greenwich Mean Time tells the user when conditions portrayed on the chart were occurring.

ISOBARS

Isobars are solid lines depicting the sea level pressure pattern. They are usually spaced at 4 millibar intervals. When the pressure gradient is weak, dashed isobars are sometimes inserted at 2 millibar intervals to more clearly define the pressure pattern. Each isobar is labelled by a two-digit number. For example, 32 signifies 1032.0 mb, 00 signifies 1000.0 mb, 92 signifies 992.0 mb, and 88 signifies 988.0 mb.

PRESSURE SYSTEMS

The letter "L" denotes a low pressure center and an "H" denotes a high pressure center. The pressure at each center is indicated by a two-digit underlined number which is interpreted the same as the isobar labels.

FRONTS

The analysis shows frontal positions and types of fronts by the symbols in figure 5-2. The "pips" on the front indicate the type of front and point in the direction toward which the front is moving. Pips on either side of a front suggest little or no movement, i.e. a stationary front. Briefing offices sometimes color the symbols to facilitate use of the map.

A three-digit number entered along a front classifies the front as to type, table 5-1; intensity, table 5-2; and character, table 5-3. For example in figure 5-1, the front extending from North Dakota southwest toward Arizona is labeled "427" which means a cold front at the surface ("4" in table 5-1), weak with little or no change ("2" in table 5-2) and with waves along the front ("7" in table 5-3). The waves along the front may be weak low pressure centers which are not indicated or simply one part of the front moving faster than the other. The triangular pips also identify this front as a cold front. The pips point toward the east over the Dakotas indicating the cold front is moving to the east in this region while in Arizona the pips point toward the southeast indicating the cold front is moving to the southeast in that region.

Two short lines across a front indicate a change in classification. Note in figure 5-1 the two lines crossing the front off the coast of Georgia indicating a change from "225" to "420". In this case a warm front extends westward from the two short lines intersecting the front while a cold front extends eastward. Note that the stationary front along the Gulf Coast is undergoing frontolysis (dissipation).

TABLE 5-1. Type of front

Code Figure	Description
0	Quasi-stationary at surface
1	Quasi-stationary above surface
2	Warm front at surface
3	Warm front above surface
4	Cold front at surface
5	Cold front above surface
6	Occlusion
7	Instability line
8	Intertropical front
9	Covergence line

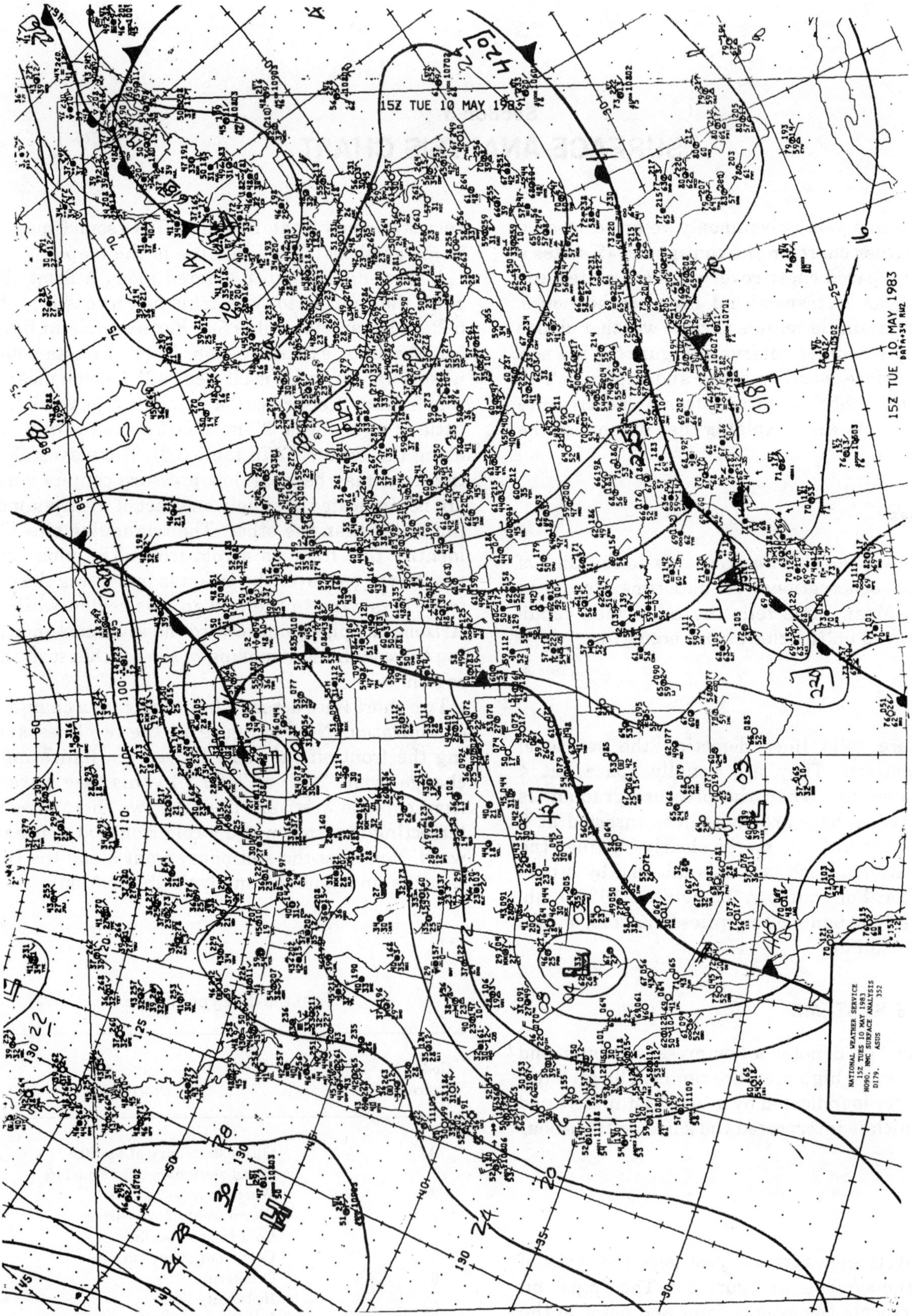

FIGURE 5-1. Surface Weather Analysis Chart.

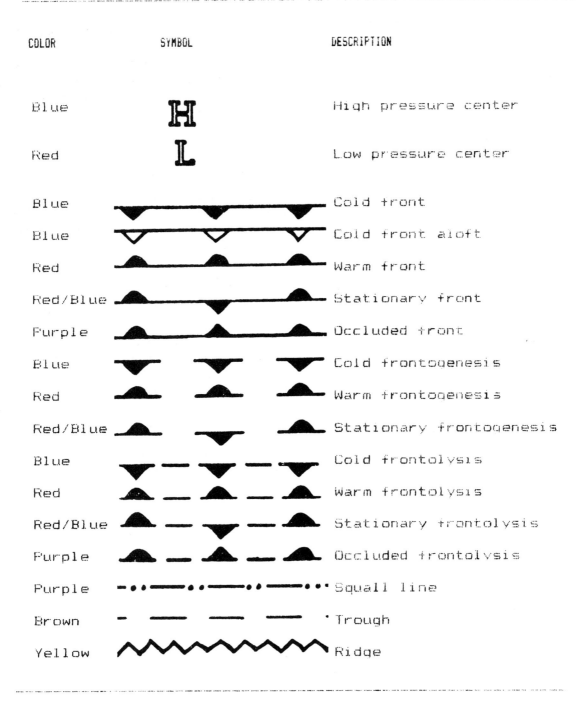

FIGURE 5-2. List of symbols on surface analyses. Colors are those suggested for on-station use.
NOTE: A trough line usually is further identified by the word "TOF".
A trough line is not a front.

TABLE 5-2. Intensity of front

Code Figure	Description
0	No specification
1	Weak, decreasing
2	Weak, little or no change
3	Weak, increasing
4	Moderate, decreasing
5	Moderate, little or no change
6	Moderate, increasing
7	Strong, decreasing
8	Strong, little or no change
9	Strong, increasing

TABLE 5-3. Character of front

Code Figure	Description
0	No specification
1	Frontal area activity, decreasing
2	Frontal area activity, little change
3	Frontal area activity, increasing
4	Intertropical
5	Forming or existence expected
6	Quasi-stationary
7	With waves
8	Diffuse
9	Position doubtful

TROUGHS AND RIDGES

A trough of low pressure with significant weather will be depicted as a thick, dashed line running through the center of the trough and identified with the word "TROF". The symbol for a ridge of high pressure is very rarely, if at all, depicted (see figure 5-2 for symbols).

OTHER INFORMATION

Figure 5-3 shows a station model which shows where the weather information is plotted. Figure 5-4 through figure 5-7 help explain the decoding of the station model.

USING THE CHART

The surface analysis provides you with a ready means of locating pressure systems and fronts. It also gives you an overview of winds, temperatures, and dew point temperatures as of chart time. When using the chart, keep in mind that weather moves and conditions change. For example, a front located over Kansas may be nearing Oklahoma by the time you see the chart. Using the surface analysis chart in conjunction with other charts such as weather depiction, radar summary, upper air, and prognostics (forecast charts) gives a more complete weather picture.

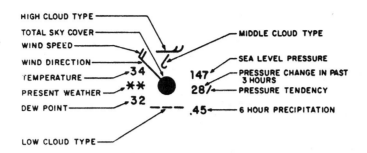

1. Total sky cover: overcast (see figure 5-4).
2. Temperature: 34 degrees F, Dew point: 32 degrees F.
3. Wind: blowing from the northwest at 20 knots relative to true north.
   Wind speeds are in knots and are indicated as:

    calm   5 knots (kts)   10 kts   15 kts   50 kts   65 kts

Circle around the station means calm. A half-flag has a value of 5 knots, a full-flag 10 knots, and a pennant 50 knots. These are used in an appropriate combination to represent wind speed; e.g. two pennants indicate wind speed of 100 knots. Wind direction stated is the direction FROM which the wind is blowing. For example:

    northwest   west   northeast   north   south

4. Present weather: continuous light snow (see figure 5-5).
5. Predominant low, mid, high cloud reported: fractostratus or fractocumulus of bad weather, altocumulus in patches, dense cirrus (see figure 5-7).
6. Sea level pressure: 1014.7 millibars (mb). Always shown as 3 digits to the nearest tenth of a millibar. For 1000 mb or greater, prefix a 10 to the 3 digits. For less than 1000 mb prefix a 9 to the 3 digits.

    108 = 1010.8 mb   888 = 988.8 mb
    225 = 1022.5 mb   961 = 996.1 mb
    000 = 1000.0 mb   720 = 972.0 mb

7. Pressure change in past 3 hours: increased steadily or unsteadily by 2.8 mb. Actual change is in tenths of a millibar. See figure 5-6 for tendency explanation.
8. 6 hour precipitation: 45 hundredths of an inch. Amount is given to the nearest hundredth of an inch.

FIGURE 5-3. Station model and explanation.

Symbol	Total sky cover
 ○	Sky clear
 ◐ (1/10 shaded vertical)	Less than 1/10 (Few)
 ◔	1/10 to 5/10 inclusive (Scattered)
◕	6/10 to 9/10 inclusive (Broken)
 ◐	10/10 with breaks (BINOVC)
●	10/10 (Overcast)
 ⊗	Sky obscured or partially obscured

FIGURE 5-4. Sky cover symbols.

WW PRESENT WEATHER (Descriptions Abridged from W. M. O. Code)

	0	1	2	3	4	5	6	7	8	9
00	Cloud development NOT observed or NOT observable during past hour	Clouds generally dissolving or becoming less developed during past hour	State of sky on the whole unchanged during past hour	Clouds generally forming or developing during past hour	Visibility reduced by smoke	Haze	Widespread dust in suspension in the air, NOT raised by wind, at time of observation	Dust or sand raised by wind, at time of observation	Well developed dust devil(s) within past hour	Duststorm or sandstorm within sight of or at station during past hour
10	Light fog	Patches of shallow fog at station, NOT deeper than 6 feet on land	More or less continuous shallow fog at station, NOT deeper than 6 feet on land		Precipitation within sight, but NOT reaching the ground	Precipitation within sight, reaching the ground, but distant from station	Precipitation within sight, reaching the ground, near to but NOT at station	Thunder heard, but no precipitation at the station	Squall(s) within sight during past hour	Funnel cloud(s) within sight during past hour
20	Drizzle (NOT freezing and NOT falling as showers) during past hour, but NOT at time of observation	Rain (NOT freezing and NOT falling as showers) during past hour, but NOT at time of observation	Snow (NOT falling as showers) during past hour, but NOT at time of observation	Rain and snow (NOT falling as showers) during past hour, but NOT at time of observation	Freezing drizzle or freezing rain (NOT falling as showers) during past hour, but NOT at time of observation	Showers of rain during past hour, but NOT at time of observation	Showers of snow, or of rain and snow, during past hour, but NOT at time of observation	Showers of hail, or of hail and rain, during past hour, but NOT at time of observation	Fog during past hour, but NOT at time of observation	Thunderstorm (with or without precipitation) during past hour, but NOT at time of observation
30	Slight or moderate dust storm or sand storm, has decreased during past hour	Slight or moderate dust storm or sand storm, no appreciable change during past hour	Slight or moderate dust storm or sand storm, has increased during past hour	Severe dust storm or sand storm, has decreased during past hour	Severe dust storm or sand storm, no appreciable change during past hour	Severe dust storm or sand storm, has increased during past hour	Slight or moderate drifting snow, generally low	Heavy drifting snow, generally low	Slight or moderate drifting snow, generally high	Heavy drifting snow, generally high
40	Fog at distance at time of observation, but NOT at station during past hour	Fog in patches	Fog, sky discernible, has become thinner during past hour	Fog, sky NOT discernible, has become thinner during past hour	Fog, sky discernible, no appreciable change during past hour	Fog, sky NOT discernible, no appreciable change during past hour	Fog, sky discernible, has begun or become thicker during past hour	Fog, sky NOT discernible, has begun or become thicker during past hour	Fog, depositing rime, sky discernible	Fog, depositing rime, sky NOT discernible
50	Intermittent drizzle (NOT freezing) slight at time of observation	Continuous drizzle (NOT freezing) slight at time of observation	Intermittent drizzle (NOT freezing) moderate at time of observation	Continuous drizzle (NOT freezing), moderate at time of observation	Intermittent drizzle (NOT freezing), thick at time of observation	Continuous drizzle (NOT freezing), thick at time of observation	Slight freezing drizzle	Moderate or thick freezing drizzle	Drizzle and rain, slight	Drizzle and rain, moderate or heavy
60	Intermittent rain (NOT freezing), slight at time of observation	Continuous rain (NOT freezing), slight at time of observation	Intermittent rain (NOT freezing) moderate at time of observation	Continuous rain (NOT freezing), moderate at time of observation	Intermittent rain (NOT freezing), heavy at time of observation	Continuous rain (NOT freezing), heavy at time of observation	Slight freezing rain	Moderate or heavy freezing rain	Rain or drizzle and snow, slight	Rain or drizzle and snow, moderate or heavy
70	Intermittent fall of snowflakes, slight at time of observation	Continuous fall of snowflakes, slight at time of observation	Intermittent fall of snowflakes, moderate at time of observation	Continuous fall of snowflakes, moderate at time of observation	Intermittent fall of snowflakes, heavy at time of observation	Continuous fall of snowflakes, heavy at time of observation	Ice needles (with or without fog)	Granular snow (with or without fog)	Isolated starlike snow crystals (with or without fog)	Ice pellets (sleet, U. S. definition)
80	Slight rain shower(s)	Moderate or heavy rain shower(s)	Violent rain shower(s)	Slight shower(s) of rain and snow mixed	Moderate or heavy shower(s) of rain and snow mixed	Slight snow shower(s)	Moderate or heavy snow shower(s)	Slight shower(s) of soft or small hail with or without rain or rain and snow mixed	Moderate or heavy shower(s) of soft or small hail with or without rain, or rain and snow mixed	Slight shower(s) of hail, with or without rain or rain and snow mixed, not associated with thunder
90	Moderate or heavy shower(s) of hail, with or without rain or rain and snow mixed, not associated with thunder	Slight rain at time of observation, thunderstorm during past hour, but not at time of observation	Moderate or heavy rain at time of observation, thunderstorm during past hour, but NOT at time of observation	Slight snow or rain and snow mixed or hail at time of observation, thunderstorm during past hour, but not at time of observation	Moderate or heavy snow, or rain and snow mixed or hail at time of observation, thunderstorm during past hour, but NOT at time of obs.	Slight or moderate thunderstorm without hail, but with rain and/or snow at time of observation	Slight or moderate thunderstorm, with hail at time of observation	Heavy thunderstorm, without hail, but with rain and/or snow at time of observation	Thunderstorm combined with dust storm or sand storm at time of observation	Heavy thunderstorm with hail at time of observation

FIGURE 5-5. Present weather.

Description of Characteristic			Description of Characteristic		
Primary Unqualified Requirement	Additional Requirements	Graphic	Primary Unqualified Requirement	Additional Requirements	Graphic
HIGHER Atmospheric pressure now higher than 3 hours ago	Increasing then decreasing	∧	THE SAME Atmospheric pressure now same as 3 hours ago	Increasing then decreasing	 ∧
	Increasing then steady; or increasing then increasing more slowly	⌐		Steady	—
				Decreasing then increasing	∨
	Steadily Increasing Unsteadily	/	LOWER Atmospheric pressure now lower than 3 hours ago	Decreasing then increasing	∨
	Decreasing or steady then increasing; or increasing then increasing more rapidly	✓		Decreasing then steady or decreasing then decreasing more slowly	 ⌐
				Steadily Decreasing Unsteadily	\
				Steady or increasing then decreasing; or decreasing then decreasing more rapidly	 ∧

FIGURE 5-6. Barometer tendencies.

CLOUD ABBREVIATION	C_L		DESCRIPTION (Abridged From W.M.O. Code)	C_M		DESCRIPTION (Abridged From W.M.O. Code)		C_H	DESCRIPTION (Abridged From W.M.O. Code)
St or Fs – Stratus or Fractostratus	1	◠	Cu of fairweather, little vertical development and seemingly flattened	1	∠	Thin As (most of cloud layer semi-transparent)	1	⌐	Filaments of Ci, or "mares tail," scattered and not increasing
Ci – Cirrus	2	◠	Cu of considerable development, generally towering, with or without other Cu or Sc bases all at same level	2	∠	Thick As, greater part sufficiently dense to hide sun (or moon), or Ns	2	⌐⌐	Dense Ci in patches or twisted sheaves, usually not increasing, sometimes like remains of Cb; or towers or tufts
Cs – Cirrostratus	3	⌒	Cb with tops lacking clear-cut outlines, but distinctly not cirriform or anvil-shaped; with or without Cu, Sc, or St	3	3	Thin Ac, mostly semi-transparent; cloud elements not changing much and at a single level	3	⌐	Dense Ci, often anvil-shaped, derived from or associated with Cb
Cc – Cirrocumulus	4	⌐	Sc formed by spreading out of Cu; Cu often present also	4	⌒	Thin Ac in patches; cloud elements continually changing and/or occurring at more than one level	4	⌐	Ci, often hook-shaped, gradually spreading over the sky and usually thickening as a whole
Ac – Altocumulus	5	⌐	Sc not formed by spreading out of Cu	5	⌐	Thin Ac in bands or in a layer gradually spreading over sky and usually thickening as a whole	5	2	Ci and Cs, often in converging bands, or Cs alone; generally overspreading and growing denser; the continuous layer not reaching 45° altitude
As – Altostratus	6	—	St or Fs or both, but no Fs of bad weather	6	⋈	Ac formed by the spreading out of Cu	6	2	Ci and Cs, often in covering bands, or Cs alone; generally overspreading and growing denser; the continuous layer exceeding 45° altitude
Sc – Stratocumulus	7	- - -	Fs and/or Fc of bad weather (scud)	7	⌐	Double-layered Ac, or a thick layer of Ac, not increasing; or Ac with As and/or Ns	7	2⌐	Veil of Cs covering the entire sky
Cu or Fc – Cumulus or Fractocumulus	8	⌐⌐	Cu and Sc (not formed by spreading out of Cu) with bases at different levels	8	M	Ac in the form of Cu-shaped tufts or Ac with turrets	8	⌐	Cs not increasing and not covering entire sky
Cb – Cumulonimbus	9	⌐⌐	Cb having a clearly fibrous (cirriform) top, often anvil-shaped, with or without Cu, Sc, St, or scud	9	⌐	Ac of a chaotic sky, usually at different levels; patches of dense Ci are usually present also	9	⌐	Cc alone or Cc with some Ci or Cs, but the Cc being the main cirriform cloud

FIGURE 5-7. Cloud abbreviation.

5-9

Section 6
WEATHER DEPICTION CHART (observed)

The weather depiction chart, figure 6-1, is computer prepared from surface aviation (SA) reports to give a broad overview of observed flying category conditions as of the valid time of the chart. The computer prepared chart is valid at the time of the plotted data. Beginning at 01Z each day, charts are transmitted at 3 hour intervals.

PLOTTED DATA

The plotted data shown for each station as required are:

Total Sky Cover

Total sky cover is shown by the station circle shaded as in table 6-1.

TABLE 6-1. Total sky cover.

Symbol	Total sky cover
○	Sky clear
⊖	Less than 1/10 (Few)
◔	1/10 to 5/10 inclusive (Scattered)
◑	6/10 to 9/10 inclusive (Broken)
◐	10/10 with breaks (BINOVC)
●	10/10 (Overcast)
⊗	Sky obscured or partially obscured

Cloud Height or Ceiling

Cloud height, above ground level, is entered under the station circle in hundreds of feet, the same as coded in a SA report. If total sky cover is few or scattered, the cloud height entered is the base of the lowest layer. If total sky cover is broken or greater, the cloud height entered is the ceiling. Broken or greater total sky cover without a height entry indicates thin sky cover. A partially or totally obscured sky is shown by the same sky cover symbol (X). However, a partially obscured sky without a cloud layer above is denoted by the absence of a height entry while a partially obscured sky with clouds above will have a cloud layer or ceiling height entry. A totally obscured sky always has a height entry of the ceiling (vertical visibility into the obscuration).

Weather and Obstructions to Vision

Weather and obstructions to vision symbols are entered just to the left of the station circle. Figure 5-5 explains most of the symbols used. When an SA Reports clouds topping ridges a symbol unique to the weather depiction chart is entered to the left of the station circle:

 denotes clouds topping ridges.

When several types of weather and/or symbols obstructions are reported at a station, only the most significant one is entered (i.e. the highest coded number in figure 5-5).

Visibility

When visibility is 6 miles or less, it is entered to the left of weather or obstructions to vision. Visibility is entered in statute miles and fractions of a mile.

Table 6-2 shows examples of plotted data.

ANALYSIS

The chart shows observed ceiling and visibility by categories as follow:

1. IFR—Ceiling less than 1,000 feet and/or visibility less than 3 miles; hatched area outlined by a smooth line.
2. MVFR (Marginal VFR)—Ceiling 1,000 feet to 3,000 feet inclusive and/or visibility 3 to 5 miles inclusive; non-hatched area outlined by a smooth line.
3. VFR—Ceiling greater than 3,000 feet or unlimited and visibility greater than 5 miles; not outlined.

The three (3) categories are also explained in the lower right portion of the chart for quick reference. Referring to figure 6-1, the MVFR conditions in southwest Oregon are indicated in an area where

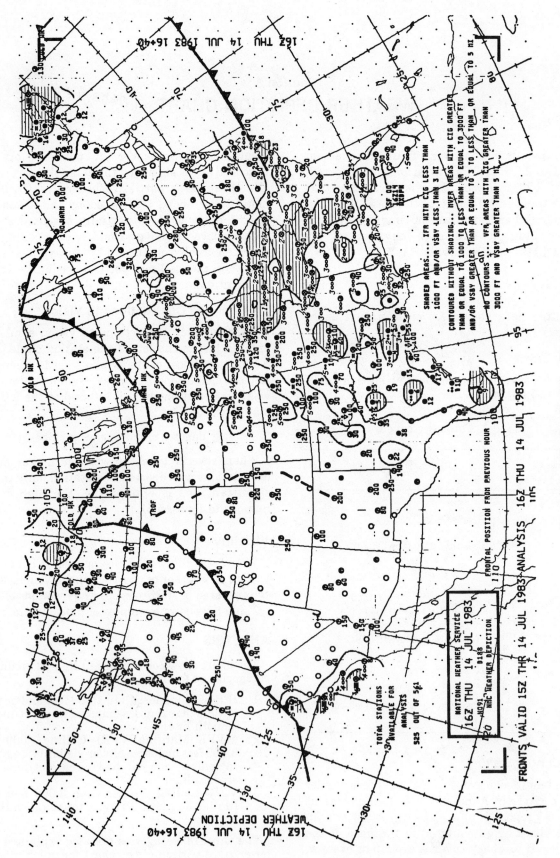

FIGURE 6-1. A Weather Depiction Chart.

TABLE 6-2. Examples of plotting on the Weather Depiction Chart

Plotted	Interpreted
① 8	Few clouds, base 800 feet, visibility more than 6
▽̇ ● 12	Broken sky cover, ceiling 1,200 feet, rain shower, visibility more than 6
5 ∞ ◐	Thin overcast with breaks, visibility 5 in haze
▲ ◑ 30	Scattered at 3,000 feet, clouds topping ridges, visibility more than 6
2 ≡ ○	Sky clear, visibility 2, ground fog or fog
½ ↑ ⊗	Sky partially obscured, visibility 1/2, blowing snow, no cloud layers observed
2 ≡ ⊗ 200	Sky partially obscured, visibility 2, fog, cloud layer at 20,000 feet. Assume sky is partially obscured since 20,000 feet cannot be vertical visibility into fog. It is questionable if 20,000 feet is lowest scattered layer or ceiling.*
¼ ✶ ⊗ 5	Sky obscured, ceiling 500, visibility 1/4, snow
1 ℞̇ ● 12	Overcast, ceiling 1,200 feet thunderstorm, rain shower, visibility 1
Ⓜ	Data missing

* Note: Since a partial and a total obscuration (X) is entered as total sky cover, it can be difficult to determine if a height entry is a cloud layer above a partial obscuration or vertical visibility into a total obscuration. Check the SA.

the plotted stations show only VFR conditions. Note that off the Baja California coast it is stated that the total stations analyzed for this chart are far more numerous than the number of stations actually plotted. Thus, there are stations in southwest Oregon, not plotted on the chart, that are reporting MVFR conditions.

In addition the chart shows fronts and troughs from the surface analysis for the preceding hour. These features are depicted the same as the surface chart.

USING THE CHART

The weather depiction chart is a choice place to begin your weather briefing and flight planning. From it, you can determine general weather conditions more readily than any other source. It gives you a "bird's eye" view at chart time of areas favorable and adverse weather and frontal systems associated with the weather.

The chart may not completely represent enroute conditions because of variations in terrain and weather between stations. Futhermore, weather changes and by the time the chart is available, plotted data around the stations have been superseded by SA reports. After you initially size up the general picture, your final flight planning must consider forecasts, progs, and the latest pilot, radar and surface weather reports.

Section 7
RADAR SUMMARY CHART (observed)

A radar summary chart, figure 7-1, graphically displays a collection of radar reports. Figure 1-2 depicts the National Weather Service radar network. The computer generated chart is valid at the time of the plotted radar reports, i.e., at H+35. Charts are available for 16 hours daily via NAFAX and 24 hours daily by DIFAX constructed from regularly scheduled radar observations. The chart displays the type of precipitation echoes and indicates their intensity, intensity trend, configuration, coverage, echo tops and bases, and movement. Severe weather watches are also plotted if they are in effect when the chart is valid. This section explains chart notations, symbols, and use.

ECHO TYPE, INTENSITY, AND INTENSITY TREND

Radar primarily detects particles of precipitation size within a cloud or falling from a cloud. The type of precipitation can be determined by the radar operator from the scope presentation in combination with other sources. TABLE 7-1 lists the symbols used to denote types of precipitation, intensity and intensity trend. The intensity is obtained from the Video Integrator Processor (VIP) and is indicated on the chart by *contours*. The six (6) VIP levels are combined into three (3) contours as indicated in TABLE 7-1. For example in Table 7-1, the area of precipitation between the first (outer) contour and the second contour would have an intensity of VIP level 1 and possibly 2. Whether we really have VIP level 2 in the area cannot be determined. However, we can say that the maximum intensity is definitely below VIP level 3. When determining intensity levels from the radar summary chart, it is recommended that the maximum possible intensity be used. To determine the actual maximum VIP level, you would examine the RAREP message (SD) for that particular time. The intensity trend is indicated by a symbol plotted beside the precipitation type. The absence of a trend symbol indicates no change. For example in figure 7-1, over eastern North Dakota and western Minnesota, there is an area of light to moderate rainshowers with no change in intensity from the previous observation. Actual intensity for frozen precipitation cannot be determined from the contours since the intensity levels are only correlated to liquid precipitation. Intensity trend for frozen precipitation is not reported on a RAREP and thus, is not indicated on the radar chart. The symbol "S+" on the chart means the area of snow is *new*. The symbol "S" on the chart means an area of snow is indicated by radar with no reference to intensity trend. Remember the intensity trend symbols; (−) decreasing, (no symbol) no changes, (+) increasing; refer only to liquid precipitation.

It is important to remember that intensity on the radar summary chart is shown by contours and *not* by the symbol following the type of precipitation. For example in figure 7-1, along the east central coast of Florida near Vero Beach, there is an area of light to moderate rainshowers that is either new or has increased in intensity from the previous observation.

Also note that hail possibly reaching the surface is associated with the thunderstorms and intense to extreme rainshowers in the southern half of Missouri. The actual locations of hail are indicated by a line drawn from "HAIL" to the symbol ■.

ECHO CONFIGURATION AND COVERAGE

The configuration is the arrangement of echoes. There are three designated arrangements, (1) a LINE of echoes, (2) an AREA of echoes and (3) an isolated CELL. A fine line would appear on the chart as a pencil thin line with a movement indicated. See section 3 under radar reports for definitions of the above.

Coverage is simply the area covered by echoes. All of the hatched area inside of the contours on the chart is considered to be covered by echoes. When the echoes are reported as a LINE, a line will be drawn through them on the radar chart. When there is at least 8/10 coverage in the line, it is labeled solid (SLD) at both ends of the line. In the absence of this label it can be assumed that there is less than 8/10 coverage. For example in figure 7-1, there is a solid line of thunderstorms with intense to extreme rainshowers extending from northeast Arkansas to southwest Kentucky.

7-1

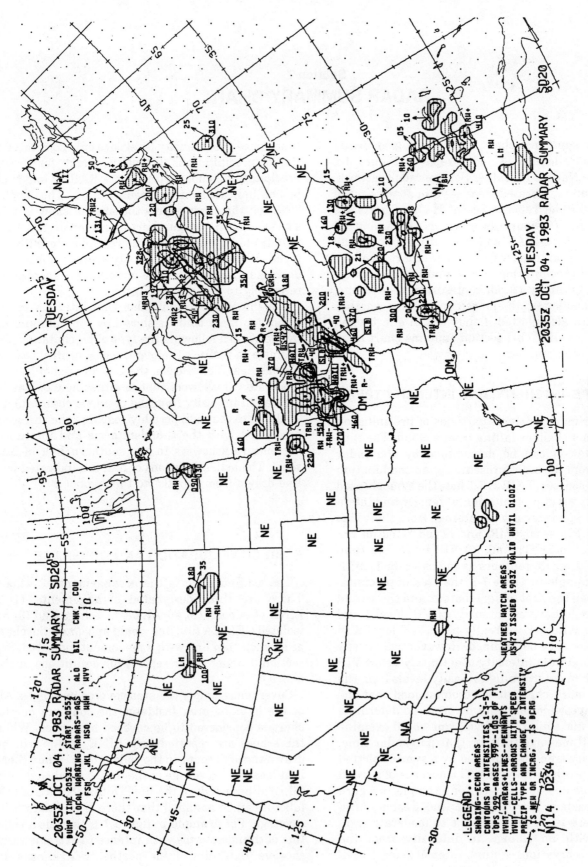

FIGURE 7-1. A Radar Summary Chart.

VIP LEVEL	ECHO INTENSITY	PRECIPITATION INTENSITY	RAINFALL RATE in/hr STRATIFORM	RAINFALL RATE in/hr CONVECTIVE
1	WEAK	LIGHT	LESS THAN 0.1	LESS THAN 0.2
2	MODERATE	MODERATE	0.1 - 0.5	0.2 - 1.1
3	STRONG	HEAVY	0.5 - 1.0	1.1 - 2.2
4	VERY STRONG	VERY HEAVY	1.0 - 2.0	2.2 - 4.5
5	INTENSE	INTENSE	2.0 - 5.0	4.5 - 7.1
6	EXTREME	EXTREME	MORE THAN 5.0	MORE THAN 7.1

* The numbers representing the intensity level do not appear on the chart. Beginning from the first contour line, bordering the area, the intensity level is 1-2; second contour is 3-4; and third contour is 5-6.

Highest precipitation top in area in hundreds of feet MSL. (45,000 FEET MSL)

SYMBOLS USED ON CHART

SYMBOL MEANING

+   INTENSITY INCREASING OR NEW ECHO
−   INTENSITY DECREASING
NO SYMBOL   NO CHANGE IN INTENSITY
35   CELL MOVEMENT TO NE AT 35 KNOTS
↱   LINE OR AREA MOVEMENT TO EAST AT 20 KNOTS
LM   LITTLE MOVEMENT
MA   ECHOES MOSTLY ALOFT
PA   ECHOES PARTLY ALOFT

SYMBOL MEANING

⟶   LINE OF ECHOES
SLD   8/10 OR GREATER COVERAGE IN A LINE
WS999   SEVERE THUNDERSTORM WATCH
WT999   TORNADO WATCH
LEWP   LINE ECHO WAVE PATTERN
HOOK   HOOK ECHO

SYMBOL MEANING

R   RAIN
RW   RAIN SHOWER
HAIL   HAIL
S   SNOW
IP   ICE PELLETS
SW   SNOW SHOWER
L   DRIZZLE
T   THUNDERSTORM
ZR, ZL   FREEZING PRECIPITATION
NE   NO ECHOES OBSERVED
NA   OBSERVATIONS UNAVAILABLE
OM   OUT FOR MAINTENANCE
STC   STC ON - all precipitation may not be seen
ROBEPS   RADAR OPERATING BELOW PERFORMANCE STANDARDS
RHINO   RANGE HEIGHT INDICATOR NOT OPERATING

RAINFALL RATES SHOULD BE USED WITH CAUTION

TABLE 7-1. Key to Radar Summary Chart.

ECHO HEIGHTS

Echo heights in locations with radars designed for weather detection are obtained by use of range height indicators and are PRECIPITATION tops and bases. In those areas not served by National Weather Service radars the tops are obtained from pilot reports and are actual CLOUD tops. Usually, echo height will be missing in the western mountain regions because ARTCC radars are used.

Heights are displayed in hundreds of feet MSL and should be considered only as approximations because of radar limitations. Tops are entered above a short line while any available bases are entered below. The top height displayed is the highest in the indicated area.

Examples are:

$\frac{220}{080}$ Bases 8,000 feet, Max top 22,000 feet

$\underline{500}$ Bases at the surface, Max top 50,000 feet

$\underline{}$
020 Bases 2,000 feet, Max top either missing or reported in another place

Absence of a figure below the line indicates that the echo base is at the surface. Radar detects tops more readily than bases because precipitation usually reaches the ground. For example in figure 7-1, over eastern North Dakota and western Minnesota the maximum precipitation top in the area is 9000 feet MSL at the location indicated by a line drawn to the symbol ■. The base of the rainshower associated with the maximum top and therefore probably over most of the area is 3000 feet MSL. This indicates dry air near the surface which is causing the rain to evaporate before reaching the ground. The level at which the precipitation is evaporating above *ground level* depends on station elevation. For example, if the ground elevation for the area just mentioned is 2500 feet MSL, the base of the precipitation is only 500 feet AGL.

ECHO MOVEMENT

Individual cell movement within a line or area is often different from that of the line or area itself. This difference is indicated by the use of different symbols, as shown in TABLE 7-1. Line or area movement is indicated by a shaft and barb combination with the shaft indicating the direction and the barbs the speed. A whole barb is 10 knots, a half barb is 5 knots, and a pennant is 50 knots. Individual cell movement is indicated by an arrow with the speed in knots entered as a number. Little movement is indentified by "LM". For example in figure 7-1, no cell movement is given over eastern North Dakota but the area movement is toward the northeast at 10 knots. Over eastern Montana, no area movement is given but the cell movement of the rainshowers is toward the east southeast at 35 knots. Over western Montana, the rainshowers show little movement.

SEVERE WEATHER WATCH AREAS

Severe weather watch areas are outlined by heavy dashed lines, usually in the form of a large rectangular box. There are two types, (1) tornado watches and (2) severe thunderstorm watches. Referring to TABLE 7-1 and figure 7-1, the type of watch and the watch number are enclosed in a small rectangle and positioned as closely as possible to the northeast corner of the watch box. For example figure 7-1, "WS 473" means a severe thunderstorm watch and is the 473rd severe weather watch issued so far in the year. The watch number is also printed at the bottom of the chart together with the issuance time and valid until time.

CANADIAN DATA

Radar data from six Canadian radar stations are plotted when available. The stations in Ontario are: Carp, Exeter, Toronto, Villeroy and Ottawa. Montreal, Quebec is also plotted. The data is displayed in AZRAN (azimuth-range) format with echo areas outlined by solid lines. Area, line and cell movements are shown in the same manner as U.S. data. An alphanumeric code associated with each echo shows, in order, area coverage, precipitation type, intensity, and intensity trend. Precipitation type and intensity trend are the same as U.S. data. For area coverage, a blank designator represents cells, a 1 equals less than 1/10 coverage, a 4 equals 1/10 to 5/10 coverage, a 7 equals 6/10 to 9/10 coverage and 10 equals 10/10 coverage. For intensity levels, 0 is very weak, 1 is weak, 2 is moderate, 3 is strong and 4 is very strong with levels 1 through 4 being comparable to the U.S. VIP LVLS 1 through 4.

For example in figure 7-1, the region in southeast Canada and covering a portion of New England constitutes a Canadian radar report "7RW2". Decoded, there is an area of moderate rainshowers with no change in intensity. 6/10 to 9/10 of the area is covered with rainshowers with area movement toward the east northeast at 20 knots. Maximum top within the area is 13,100 feet MSL.

Canadian echo top reports are converted from meters to feet and are plotted to the nearest hundreds of feet MSL. For example, west of New York State in southern Canada, the tops are decoded as follows: "197" is 19,700 feet MSL and "328" is 32,800 feet MSL.

It can sometimes be difficult to interpret the data where both American and Canadian reports are plotted, such as in the Great Lakes region in figure 7-1.

Do not confuse Canadian data with a severe weather watch box. In figure 7-1, the rectangular box in the Great Lakes region is not a severe weather watch box but a Canadian radar plot. This box is *not* outlined by heavy dashed lines as is required for a severe weather watch area. On charts transmitted after December 1983, the areas of Canadian data are plotted as light solid lines instead of heavy solid lines as shown in figure 7-1. This should lead to less confusion as to U.S. verus Canadian data.

USING THE CHART

The radar summary chart aids in preflight planning by identifying general areas and movement of precipitation and/or thunderstorms. Radar detects ONLY drops or ice particles of precipitation size, it DOES NOT detect clouds and fog. Therefore, the absence of echoes does not guarantee clear weather. Furthermore, cloud tops may be higher than precipitation tops detected by radar. The chart must be used in conjunction with other charts, reports, and forecasts.

Examine chart notations carefully. Always determine location and movement of echoes. If echoes are anticipated near your planned route, take special note of echo intensity and trend. Be sure to examine for missing radar reports (NA, OM) before briefing "no echoes present". For example, the Covington (CVG) radar report in northern Kentucky is shown as not available (NA). There could very well be echoes in southwest Ohio but too far away to be detected by the other surrounding radars.

Suppose your proposed route will take you through an area of widely scattered thunderstorms with no increase anticipated. When these storms are separated by good VFR weather, you most likely can pick your way among them, visually sighting and circumnavigating the storms. However, widespread cloudiness may conceal the thunderstorms. To avoid these embedded thunderstorms, you must either use airborne radar or detour the area. Most details on avoiding hazards of thunderstorms are given in Chapter 11, Aviation Weather.

Keep in mind that the chart is for preflight planning only and should be updated by hourly radar reports. Once airborne, you must evade individual storms from inflight observations either by visual sighting or by airborne radar or request weather radar echo information from FSS Flight Watch which has access to Radar Remote Weather Displays (RRWDS).

One more thought before this section ends. There can be an interpretation problem concerning an area of precipitation that is reported by more than one radar site. As an example, station A may be reporting RW− with cell movement to the northeast at 10 knots while station B may be reporting TRW+ with cell movement to the northeast at 30 knots for the same area. This difference in reports may be due to the different perspective and distance of the radar sites from the area of echoes. The area may be moving away from station A and approaching station B. The rule of thumb is to use that plotted data associated with the area which presents the greatest hazard to aviation, i.e. in the case above, the plotted data from station B. In figure 7-1, the area of rainshowers in eastern Montana should be briefed as having no change in intensity (RW) rather than decreasing in intensity (RW−).

Section 8
SIGNIFICANT WEATHER PROGNOSTICS (forecast)

Significant weather prognostic charts, called "progs" for brevity, portray forecast weather which may influence flight planning. TABLE 8-1 explains some symbols used on these charts. Significant weather progs are issued both for domestic and international flights.

TABLE 8-1. Some standard weather symbols

Symbol	Meaning	Symbol	Meaning
⌒	Moderate turbulence	▽	Rain shower
⌒⌒	Severe turbulence	❊▽	Snow shower
ᴗ	Moderate icing	⌐	Thunderstorms
ᴗᴗ	Severe icing	∿	Freezing rain
•	Rain	6	Tropical storm
✱	Snow	◉	Hurricane (typhoon)
ͻ	Drizzle		

NOTE: Character of stable precipitation is the manner in which it occurs. It may be intermittent or continuous. A single symbol denotes intermittent and a pair of symbols denotes continuous.

Examples,

Intermittent	Continuous	
•	• •	Rain
ͻ	ͻ ͻ	Drizzle
✱	✱ ✱	Snow

DOMESTIC FLIGHTS

Significant weather progs are manually prepared by the forecaster for the conterminous U.S. and adjacent areas. The U.S. low level significant weather prog is designed for domestic flight planning to 24,000 feet and a U.S. high level prog is for domestic flights from 24,000 feet to 63,000 feet. Chart legends include valid time in GMT.

U.S. Low Level Significant Weather Prog

The low level prog is a four-panel chart as shown in figure 8-1. The two lower panels are 12- and 24-hour surface progs. The two upper panels are 12- and 24-hour progs of significant weather from the surface to 400 millibars (24,000 feet). The charts show conditions as they are forecast to be at the valid time of the chart. A chart is issued four (4) times daily; the 12 and 24 hour forecasts are based on the 00Z, 06Z, 12Z and 18Z synoptic data. For example, the prog in figure 8-1 is based on the 12Z 9 NOV initial data.

Surface Prog. The two surface prog panels use standard symbols for fronts and pressure centers explained in section 5. Movement of each pressure center is indicated by an arrow showing direction and a number indicating speed in knots. Isobars depicting forecast pressure patterns are included on some 24-hour surface progs.

The surface prog also outlines areas of forecast precipitation and/or thunderstorms as shown in the lower panels of figure 8-1. Smooth lines enclose areas of expected continuous or intermittent (stable) precipitation; dash-dot lines enclose areas of showers or thunderstorms (unstable precipitation). Areas of continuous or intermittent precipitation with embedded showers and thunderstorms will also be enclosed by dash-dot lines.

Note that symbols indicate precipitation type and character (see TABLE 8-1 and 8-2). If precipitation will affect half or more of an area, that area is shaded; absence of shading denotes more sparse precipitation, specifically less than half areal coverage. Look at the lower left panel of figure 8-1. At 0000Z the forecast is for continuous snow and rain affecting half or more of an area extending from north of the

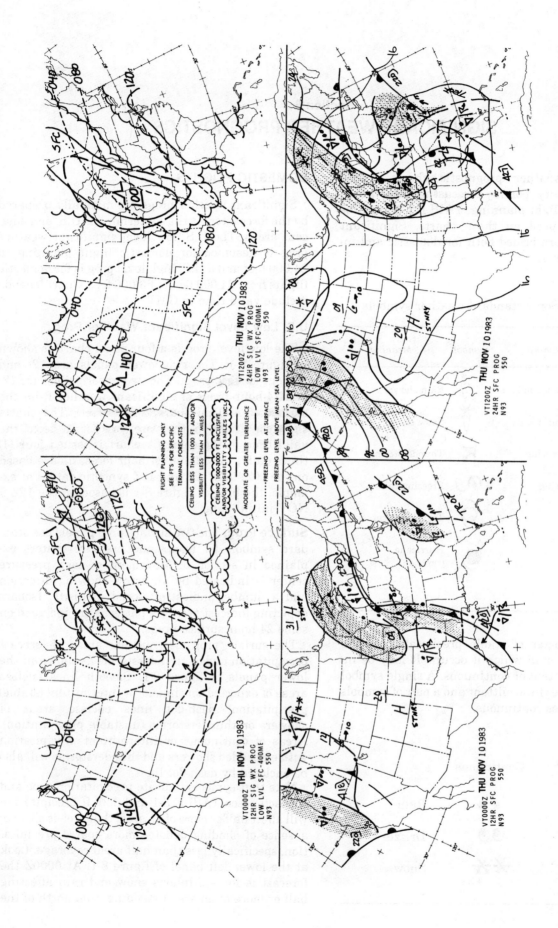

FIGURE 8-1. U.S. Low Level Significant Weather Prog (Sfc-400 mb).

TABLE 8-2. Significant weather prognostic symbols

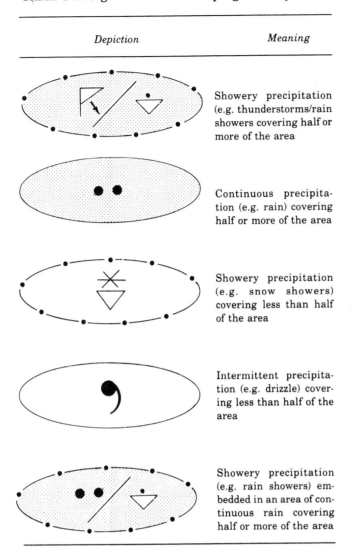

Great Lakes southwestward to Kansas. The snow is forecast over the west portion of the area with the rain-snow line represented by a dashed line. On the same prog, along the west coast and extending inland to Montana, embedded showers are forecast within an area of continuous rain. Along the coast, coverage is expected to be half or more of the area while further inland coverage is expected to be less than half of the area. From central Illinois southward to central Louisiana, showers and thunderstorms are forecast with half or more areal coverage, however, further south coverage is expected to be less than half.

Significant Weather. The upper panels of figure 8-1 depict IFR, MVFR, turbulence, and freezing levels. Note the legend near the center of the chart which explains methods of depiction.

Smooth lines enclose areas of forecast IFR weather and scalloped lines enclose areas of marginal weather (MVFR). VFR areas are not outlined. This is NOT the same manner of depiction used on the weather depiction chart to portray IFR and MVFR. Referring to figure 8-1, at 00Z an area of IFR is depicted along the east coast from North Carolina to northern Florida and is surrounded by an area of MVFR. Note that depictions are *not* extended over the open waters even though IFR conditions may exist.

Forecast areas of moderate or greater turbulence are enclosed by long-dashed lines. Thunderstorms always imply moderate or greater turbulence; thus the area of thunderstorm turbulence will not be outlined.

A symbol entered within a general area of forecast turbulence denotes intensity. Figures below and above a short line show expected base and top of the turbulent layer in hundreds of feet MSL. Absence of a figure below the line indicates turbulence from the surface upward. No figure above the line indicates turbulence extending above the upper limit of the chart. Turbulence forecast from the surface to above 24,000 feet is indicated by the notation "SFC" below the line with the upper value left blank. Referring to figure 8-1, at 00Z an area of moderate non-thunderstorm related turbulence is forecast over the far western U.S. from the surface to 14,000 feet MSL; moderate non-thunderstorm related turbulence is forecast over the middle part of the country and over eastern Maine from the surface to 12,000 feet MSL. Thunderstorm related turbulence is indicated on the lower panels by forecast areas of thunderstorms.

Freezing level height contours for the *highest* freezing level are drawn at 4,000 foot intervals. The 4,000 foot contour terminates at the 4,000 foot terrain level along the Rocky Mountains. Contours are labelled in hundreds of feet MSL. The dotted line shows where the freezing level is forecast to be at the surface and is labelled "32 F" or "SFC". An upper freezing level contour crossing the surface 32 degree line indicates multiple freezing levels due to layers of warmer air aloft. If clouds and precipitation are forecast in this area, icing hazards should be considered.

The low level significant weather prog does not specifically outline areas of icing. However, icing is implied in clouds and precipitation above the freezing level. Interpolate for freezing levels between the given contours. For example in figure 8-1, at 00Z the forecast *highest* freezing level over Oklahoma City is approximately 6,000 feet MSL.

36 and 48 Hour Surface Weather Prog

This prog is an extension of those 12 and 24 hour surface prog panels which are based on 00Z and 12Z initial synoptic data. The prog in figure 8-2 is a continuation of the 12 and 24 hour prog in figure 8-1.

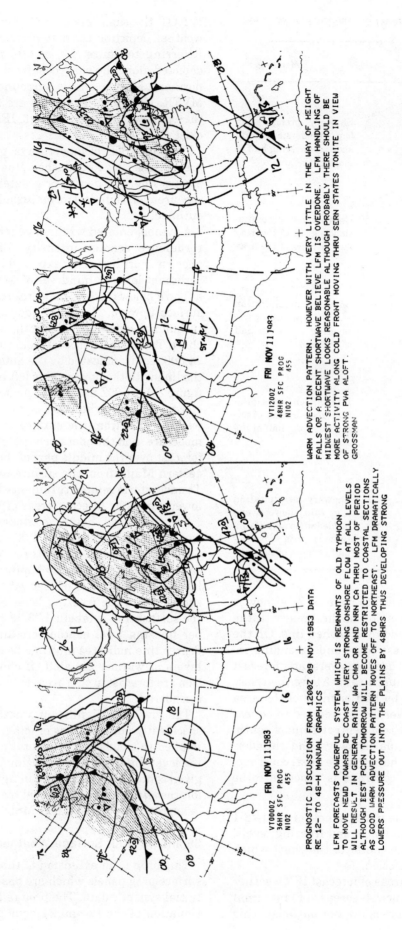

FIGURE 8-2. U.S. Low Level 36 and 48 hour Significant Weather Prog.

The depiction of data is the same as on the 12 and 24 hour surface prog with the following exceptions:

No freezing precipitation is forecast.

Scalloped lines denote area of *overcast* clouds with *no* reference to the height of cloud base.

A prognostic discussion is included to explain the forecaster's reasoning for the 12 hour through 48 hour surface progs.

Use of the Chart.

The 36 and 48 hour surface prog should only be used for outlook purposes, i.e. to get a very generalized weather picture of conditions that are in the relatively distant future.

High Level Significant Weather Progs

Manually produced forecast charts of significant weather are available for both domestic and international flights. The U.S. National Meteorological Center (NMC), near Washington, D.C., is a component of the World Area Forecast System (WAFS). NMC is designated in the WAFS as both a World Area Forecast Center and a Regional Area Forecast Center (RAFC). Its main function as a World Area Forecast Center is to prepare global forecasts in grid-point form of upper winds and upper air temperatures and to supply the forecasts to associated RAFCs. One of its main RAFC functions is to prepare and supply to users charts of forecast winds and temperatures and of forecast significant weather. This section will deal with the content of significant weather progs.

Significant weather to be depicted on the charts are the following:

a. cumulonimbus (CB) clouds meeting at least one of the following criteria:
  1. widespread cumulonimbus clouds or cumulonimbus clouds along a line with little or no space between individual clouds;
  2. cumulonimbus clouds embedded in cloud layers or concealed by haze or dust;
b. tropical cyclones;
c. severe squall lines;
d. moderate or severe turbulence (in cloud or clear air);
e. widespread sandstorm/duststorm;
f. surface positions, speed and direction of movement enroute weather phenomena;
g. tropopause heights; and
h. jetstreams.

Depiction of Thunderstorms and Cumulonimbus Clouds (CB)

Required thunderstorm activity is to be depicted by means of the abbreviation, or symbol, "CB". By definition, this symbol is to refer to the occurrence or expected occurrence of an area of widespread cumulonimbus clouds along a line with little or no space between individual clouds, or cumulonimbus clouds embedded in cloud layers or concealed by haze or dust. It does not refer to isolated or scattered (occasional) cumulonimbus clouds not embedded in cloud layers or concealed by haze or dust. The symbol "CB" automatically implies moderate or greater turbulence and icing; thus, these associated hazards will not be depicted separately.

CB data will normally be identified as ISOL EMBD CB (isolated embedded CB), OCNL EMBD CB (occasional embedded CB), ISOL CB in HAZE (isolated CB in HAZE) or OCNL CB in HAZE (occasional CB in haze). In rare instances, CB coverage above FL240 may exceed 4/8 coverage; in these instances, CB activity will be described as FRQ CB (frequent, cumulonimbus clouds with little or no separation). The meanings of these area coverage terms are: ISOL, less than 1/8; OCNL, 1/8 to 4/8; and FRQ, 5/8 to 8/8.

CB bases are considered certain to be below FL240 and will be shown as XXX. CB tops are to be expressed in hundreds of feet MSL. The area to which the forecast applies will be shown by scalloped lines. Examples:

Meaning: Occasional (1/8 to 4/8 area coverage) embedded cumulonimbus clouds with bases below FL240 and tops forecast to reach FL450.

Meaning: Isolated (less than 1/8 area coverage) embedded cumulonimbus clouds with bases below FL240 and tops forecast to reach FL350.

Depiction of Tropical Cyclones

Tropical storms are depicted by the symbol 𝟞. Areas of associated cumulonimbus activity, if meeting the previously given criteria (ISOL EMBD CB, OCNL EMBD CB, ISOL CB IN HAZE, OCNL CB IN HAZE. FRQ CB), are enclosed by scalloped lines and labelled with the vertical extent. Example:

Meaning: Forecast position of a tropical cyclone with no associated thunderstorm area.

Meaning: A thunderstorm area (5/8 to 8/8 area coverage, bases below FL240, tops FL500) associated with a tropical cyclone.

Notes: 1. The names of tropical cyclones, when relevant, will be entered adjacent to the symbol.
2. Significant weather chart depicting the tropical cyclone symbol will have a statement to the effect that the latest tropical cyclone advisory, rather than the tropical cyclone's prognostic position on the chart, is to be given public dissemination.

Depiction of Severe Squall Lines
Severe squall lines are depicted within areas of CB activity by the symbol:

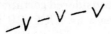

Example of depiction of severe squall line and associated CB activity:

Meaning: Forecast severe squall line with associated CB, coverage 5/8 to 8/8, bases below FL240 and tops forecast to reach FL500.

Depiction of Clear Air Turbulence
Area of forecast moderate or greater clear air turbulence (CAT) are bounded by heavy dashed lines. Clear air turbulence is interpreted as including all turbulence (including wind-shear induced and mountain-wave induced) not caused by convective activity. Areas are labelled with the appropriate symbol (⋏ for moderate CAT; ⋀ for severe CAT) and the vertical extent in hundreds of feet MSL.
Examples:

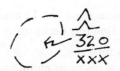

Meaning: An area of forecast moderate CAT, vertical extent from FL280 to FL360.

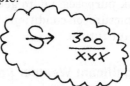

Meaning: An area of forecast severe CAT, vertical extent from below FL240 to FL320.

Note: THE SYMBOL CB IMPLIES HAIL, MODERATE OR GREATER TURBULENCE AND ICING.

Depiction of Widespread Sandstorm or Duststorm
Areas of these phenomena are enclosed by scalloped lines and labelled by symbol and vertical extent. Example:

Meaning: Widespread sandstorm or duststorm, bases below FL240 (i.e., at the surface) tops FL300.

Depiction of Inter-Tropical Convergence Zone
Areas of associated cumulonimbus activity, if meeting the previously given criteria (ISOL EMBD CB, OCNL EMBD CB, ISOL CB IN HAZE, OCNL CB IN HAZE, FRQ CB) are enclosed by scalloped lines and labelled with the vertical extent.
Example:

Meaning: Forecast position of inter-tropical convergence zone, with associated thunderstorm areas, coverage respectively frequent (5/8 to 8/8), bases below FL240, tops FL450; and occasional (1/8 to 4/8), bases below FL240, tops FL350.

Depiction of Fronts
Forecast surface position and speed (knots) and direction of movement of frontal systems associated with significant weather are depicted. Example:

Meaning: A frontal system forecast to be at the position and with the orientation indicated at the valid time of the prognostic chart. Forecast movement related to true north and speed in knots are indicated by arrow shafts and adjacent numbers.

Depiction of Tropopause Heights

Tropopause heights are to be depicted in hundreds of feet MSL as 240, 270, 300, 340, 390, 450, and 530. Other heights may be used occasionally to define areas of very flat tropopause slope. Heights depicted are enclosed in small rectangular blocks. For example, in figure 8-3, note how the tropopause slopes from 39,000 feet in southern Minnesota to 45,000 feet in Texas.

Depiction of Jetstreams

The height and maximum wind speed of jetstreams having a core speed of 80 knots or greater are shown. Height is given as flight level (FL). Points along the jetstream at which the maximum wind speed is forecast are depicted with shafts, pennants, and feathers. Example:

Meaning: A jetstream with forecast maximum speed of 100 knots at a height of 42,000 feet at another location and 90 knots at a height of 37,000 feet at another location. Wind speed along other portions is forecast to be less. Wind directions are indicated by the orientation of arrow shafts in relation to true north.

U.S. High Level Significant Weather Prog

The U.S. high level significant weather prog, figure 8-3, encompasses airspace from 24,000 feet to 63,000 feet pressure altitude. The prog is manually produced by the U.S. National Meteorological Center (NMC). Figure 8-3 outlines areas of forecast turbulence and cumulonimbus clouds. Table 8-3 interprets some examples of chart annotation.

Turbulence. Large-dashed lines enclose areas of probable moderate or greater turbulence not caused by convective activity. Symbols denote intensity, base, and top.

Cumulonimbus Clouds. Small-scalloped lines enclose areas of expected cumulonimbus development. The contraction "CB" denotes cumulonimbus. This symbol refers to the occurrence or expected occurrence of an area of widespread cumulonimbus clouds or cumulonimbus clouds along a line with little space between the individual clouds. It also depicts cumulonimbus clouds embedded in cloud layers or concealed by haze or dust. It does not refer to isolated or scattered cumulonimbus clouds not embedded in cloud layers or concealed by haze or dust. Cumulonimbus clouds *imply* moderate or greater turbulence and icing.

Cumulonimbus coverage and heights represent an overall average for the forecast area. When a wide variation is expected within an area, separate CB amounts and heights may be indicated.

The meanings of area coverage terms are: ISOL, less than 1/8; OCNL, 1/8 to 4/8; and FRQ, 5/8 to 8/8. CB bases are considered certain to be below 24,000 feet and are shown as XXX.

TABLE 8-3. Depiction of clouds and turbulence on a High Level Significant Weather Prog

	Depiction	Meaning
1.	ISOL EMBD CB 420/XXX	ISOL embedded (less than one-eight cumulonimbus, tops 42,000 feet. Bases are below 24,000 ft.—the lower limit of the prog.
2.	OCNL EMBD CB 520/XXX	1/8 to 4/8 coverage, embedded cumulonimbus, tops 52,000 feet, bases below 24,000 feet.
3.	FRQ CB 330/XXX	5/8 to 8/8 coverage cumulonimbus, bases below 24,000 feet and tops 33,000 feet.
4.	330/XXX to	Moderate to severe turbulence from below lower limit of the prog (24,000 feet) to 33,000 feet. (Consult low-level prog for turbulence forecasts below 24,000 feet.)
5.	ABV 630/350	Moderate turbulence from 35,000 feet to above upper limit of the prog.

NOTES:

Base and top shown by figures below and above a short line respectively.

Cumulonimbus Clouds, Examples 1, 2 and 3. Bases always below 24,000 feet and are shown by XXX. Tops above the upper limits of chart shown

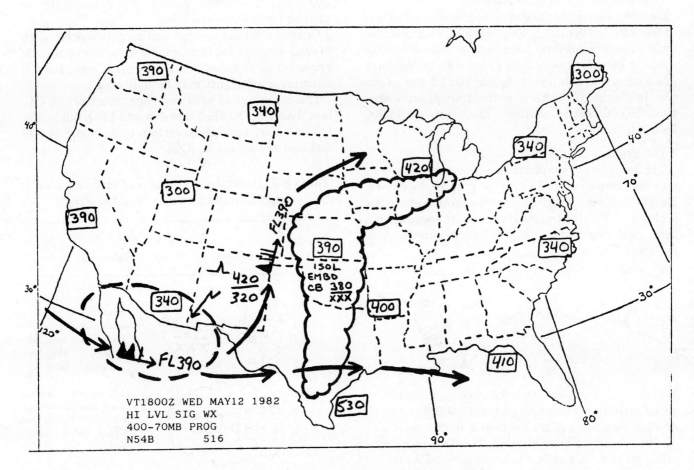

FIGURE 8-3. U.S. High Level Significant Weather Prog.

as "ABV 630" depending on the chart.
Turbulence. Bases and tops depicted the same as for cumulonimbus clouds.

International Flights

Figure 8-4 is an example of the significant weather prog for international flights. A manually produced forecast chart of high level significant weather.

Referring to figure 8-4, the legend shows NMC as a Regional Area Forecast Center (RAFC) and the originator of this significant weather prog. Significant weather is limited to the occurrence or expected occurrence of meteorological conditions considered to be of concern to aircraft operations. Significant weather progs are prepared only for flight levels from 25,000 feet to 60,000 feet. The valid time of this particular prog in figure 8-4 is 0000Z on February 9, 1984. All heights are in Flight Level (FL) and in hundreds of feet MSL.

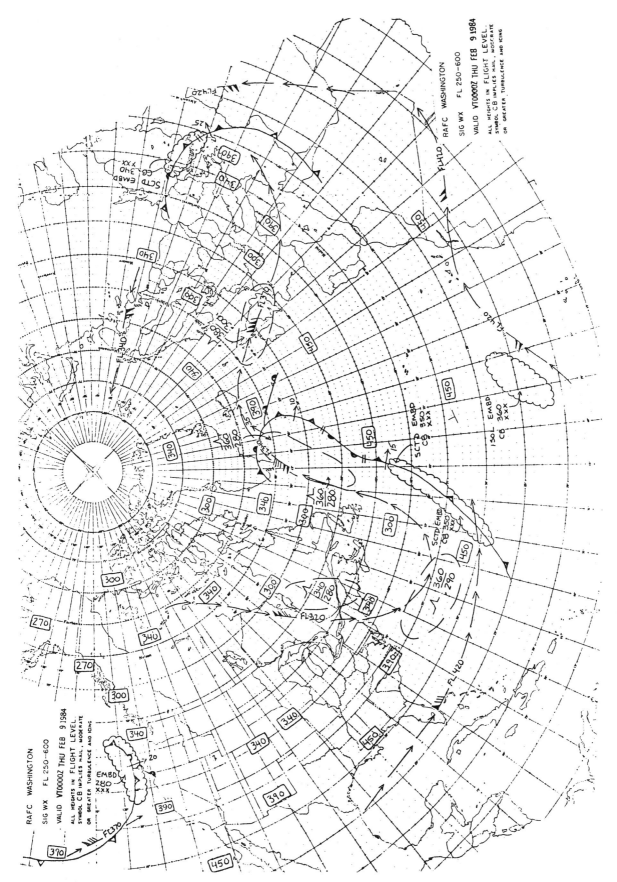

FIGURE 8-4. International High Level Significant Weather Prog Chart.

Section 9
WINDS AND TEMPERATURES ALOFT *(forecast)*

Winds aloft, both forecast and observed, are computer prepared and routinely transmitted by facsimile. The forecast winds aloft charts also contain forecast temperatures aloft.

FORECAST WINDS AND TEMPERATURES ALOFT (FD)

Forecast winds and temperatures aloft charts are prepared for eight levels on eight separate panels. Those levels being 6,000; 9,000; 12,000; 18,000; 24,000; 30,000; 34,000 and 39,000 feet MSL. They are available daily as 12-hour progs valid at 1200Z and 0000Z. A legend on each panel shows the valid time and the level of the panel. Levels below 18,000 feet are true altitudes. Levels 18,000 and above are pressure altitudes of flight levels. Figure 9-1 is one panel of a winds and temperatures aloft forecast.

Temperature in whole degrees Celsius for each forecast point is entered above the station circle. Arrows with pennants and barbs similar to those used on the surface map show wind direction and speed. Wind direction is drawn to the nearest 10 degrees with the second digit of the coded direction entered at the outer end of the arrow. First you determine the general direction to the nearest 10 degrees. For example, a wind in the northwest quadrant with a digit 3 indicates 330 degrees. A calm or light and variable wind is shown by "99" entered to the lower left of the station circle.

Following are examples of plotted temperatures and winds with their interpretations:

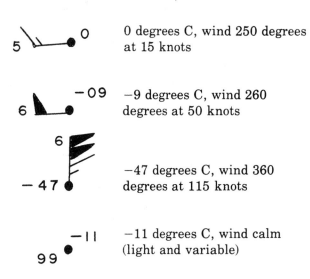

Plotted	Interpretation
12 ⟍6	12 degrees C, wind 060 degrees at 5 knots
3 ⟋6	3 degrees C, wind 160 degrees at 25 knots

- 0 degrees C, wind 250 degrees at 15 knots
- −9 degrees C, wind 260 degrees at 50 knots
- −47 degrees C, wind 360 degrees at 115 knots
- −11 degrees C, wind calm (light and variable)

OBSERVED WINDS ALOFT

Charts of observed winds for selected levels are sent twice daily on a four panel chart valid at 1200Z and 0000Z as shown in figure 9-2. Wind direction and speed at each observing station (figure 1-3) is shown by arrows the same as on the forecast charts. A calm or light and variable wind is shown as "LV" and a missing wind as "M", both plotted to the lower right of the station circle. The station circle is filled in when the reported temperature—dew point spread is 5 degrees Celsius or less. Figure 9-3 is a panel of the observed winds aloft chart. Observed temperatures are included on the upper two (2) panels (24,000 feet and 34,000 feet). A dotted bracket around the temperature means a calculated temperature.

The second standard level for a reporting station is found between 1,000 feet and 2,000 feet above the surface, depending on station elevation. To compute the second standard level, find the next thousand foot level above the station elevation and add 1,000 feet to that level. For example, the next thousand foot level above Oklahoma City OK (station elevation 1,290 feet MSL) is 2,000 feet MSL. The second standard level for Oklahoma City OK (2,000 feet + 1,000 feet) is 3,000 feet MSL or 1,710 feet AGL.

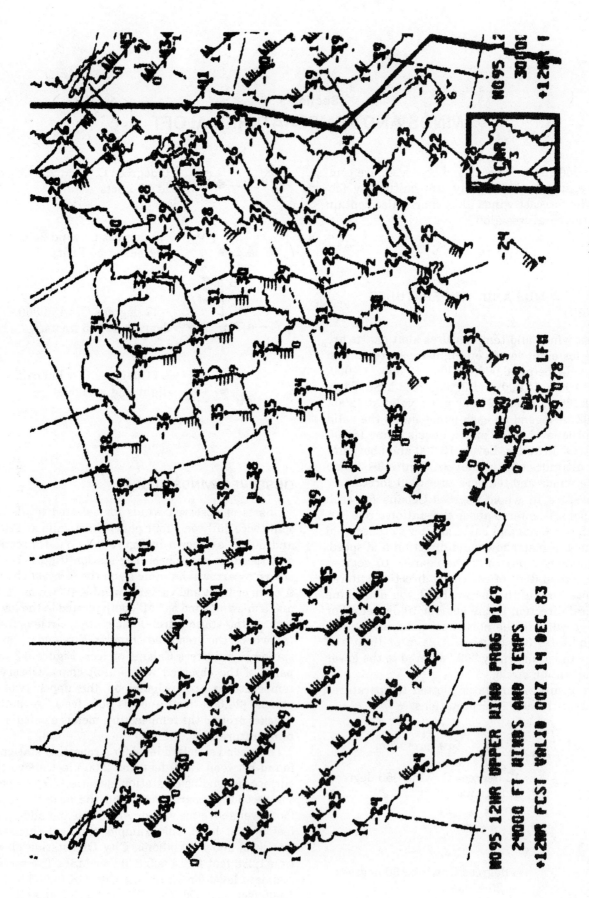

FIGURE 9-1. A panel of winds and temperatures aloft forecast for 24,000 feet pressure altitude.

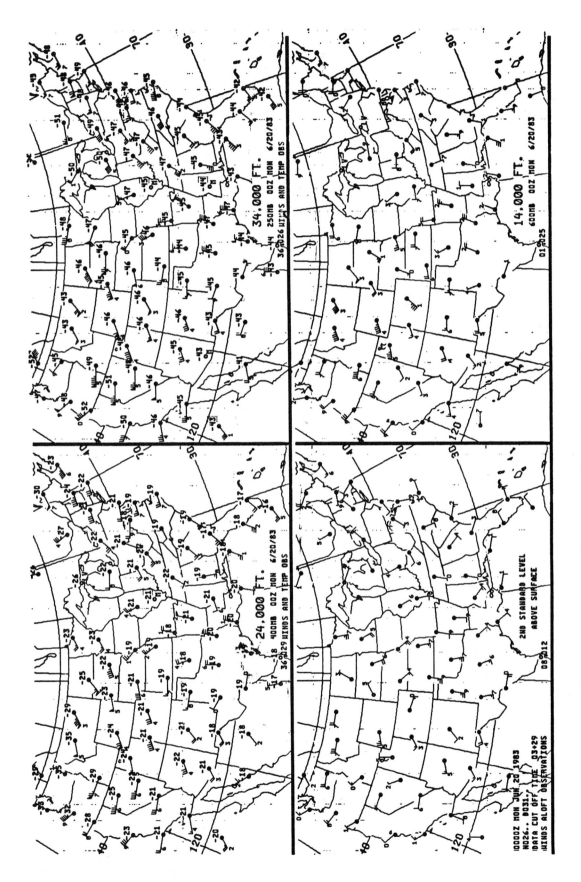

FIGURE 9-2. An Observed Winds Aloft Chart.

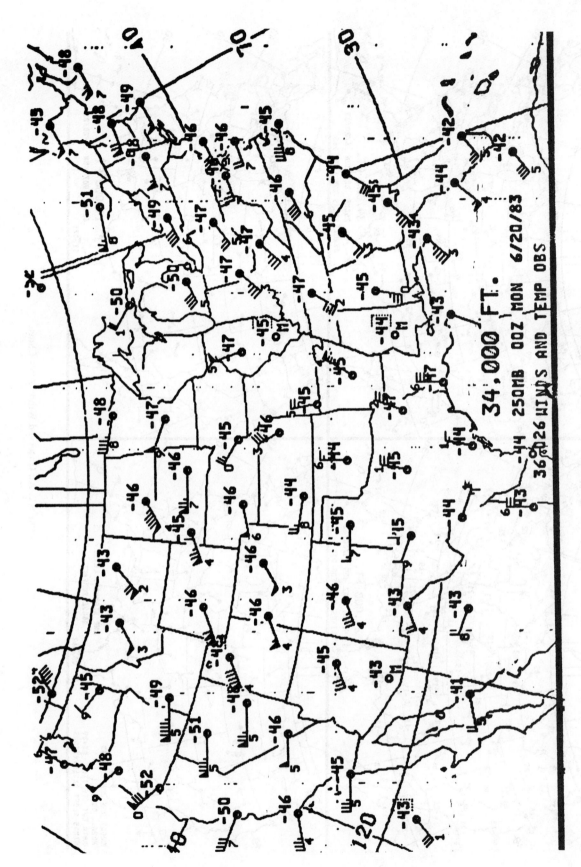

FIGURE 9-3. A panel of observed winds aloft for 34,000 feet.

Examples:

Station	Denver CO	Bismarck ND	Topeka KS	Key West FL
Station elevation:	3604 MSL	1677 MSL	879 MSL	0 MSL
Next thousand foot level above station:	4000 MSL +1000	2000 MSL +1000	1000 MSL +1000	1000 MSL +1000
Second standard level:	5000 MSL or 1396 AGL	3000 MSL or 1323 AGL	2000 MSL or 1121 AGL	2000 MSL or 2000 AGL

The 14,000 feet MSL panel is true altitude while the 24,000 and 34,000 feet MSL panels are pressure altitude.

USING THE CHARTS

The use of winds aloft charts seems obvious—to determine winds at a proposed flight altitude or to select the best altitude for a proposed flight. Temperatures also can be determined from the forecast charts. To determine winds and temperatures at a level between chart levels, interpolate. Also interpolate the data when the time period is other than the valid time of the chart.

Forecast winds are generally preferable to observed winds since they are more relevant to flight time. Although, observed winds are 5 to 8 hours old when received by facsimile and their reliability diminishes with time they can be a useful reference to check for gross errors on the 12-hour prog.

INTERNATIONAL FLIGHTS

Computer generated forecast charts of winds and temperatures aloft are available for international flights at specified levels. The U.S. National Meteorological Center (NMC), near Washington, D.C., is a component of the World Area Forecast System (WAFS). NMC is designated in the WAFS as both a World Area Forecast Center and a Regional Area Forecast Center (RAFC). Its main function as a World Area Forecast Center is to prepare global forecasts in grid-point form of upper winds and upper air temperatures and to supply the forecasts to associated RAFCs. One of NMC's main RAFC functions is to prepare and supply to users charts of forecast winds and temperatures. Figure 9-4 and 9-5 are examples of the forecast winds and temperatures aloft charts for international flights.

For example on figures 9-4 and 9-5, the lower left portion of the chart shows the originating office, NMC (Washington DC); flight level of chart (34,000 feet MSL); valid time of the chart and data base time (data from which the forecast was derived). Forecast winds are expressed in knots for spot locations with direction and speed depicted in the same manner as the U.S forecast winds and temperatures aloft chart (figure 9-1). Forecast temperatures are depicted for spot locations inside small circles; expressed in degrees Celsius. For charts with flight levels at or below FL180 (18,000 feet), temperatures are depicted as negative (−) or positive (+). On charts for flight levels (FL) above FL180, temperatures are always negative and so no sign (− or +) will be depicted.

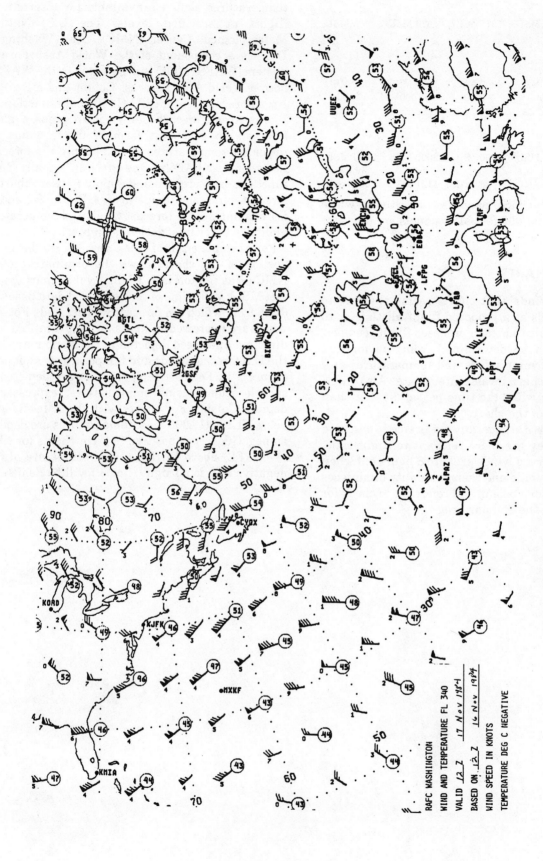

FIGURE 9-4. Polar stereographic forecast winds and temperatures aloft chart.

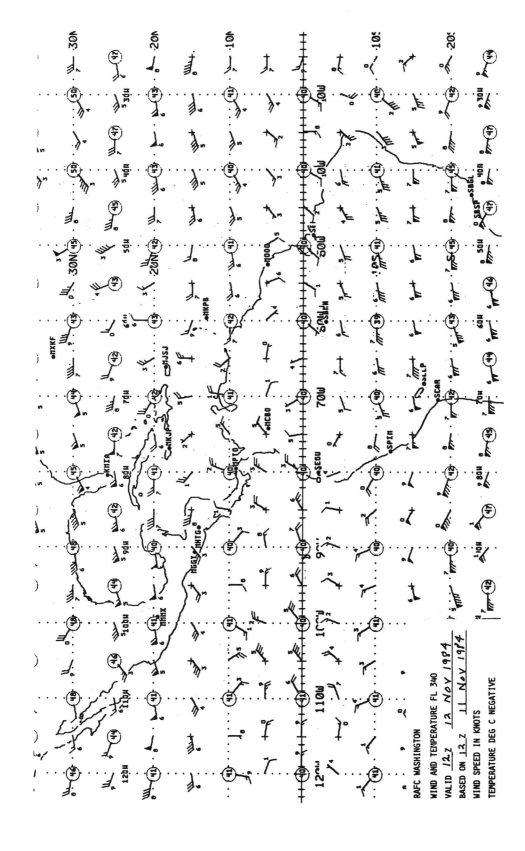

Figure 9-5. Mercator forecast winds and temperatures aloft chart.

Section 10
COMPOSITE MOISTURE STABILITY CHART (observed)

The Composite Moisture Stability Chart, figure 10-1, is an analysis chart using observed upper air data. It is composed of the following four panels: stability, freezing level, precipitable water, and average relative humidity. This computer generated chart is available twice daily with valid times of 12Z and 00Z. The availability of upper air data (on all the panels) for analysis is indicated by the shape of the station model. Use the legend on the precipitable water panel of figure 10-4 for the explanation. On this chart, the mandatory levels used are the surface, the 1000 mb, the 850 mbs, the 700 mb, and the 500 mb. Significant levels are the levels between the mandatory levels where significant changes in temperature and/or moisture occur when compared to below or above that level.

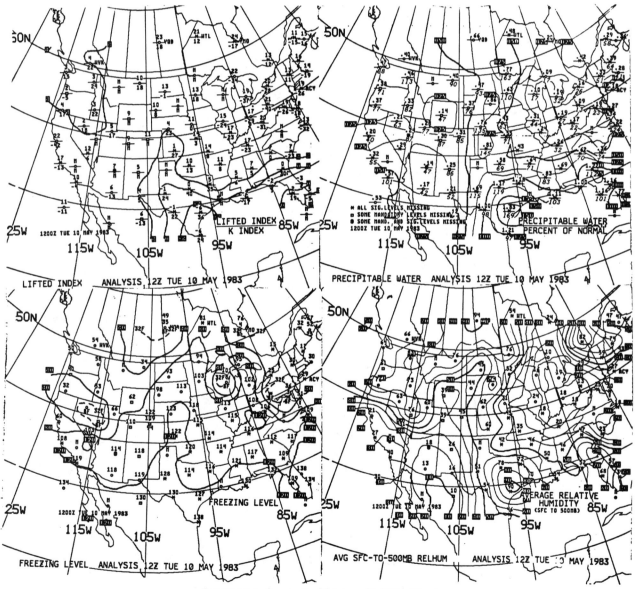

FIGURE 10-1. Composite Moisture Stability Chart.

10-1

STABILITY PANEL

The stability panel (upper left panel of chart), (figure 10-2), outlines areas of stable and unstable air. Two stability indices are computed for each upper air station, one is the *lifted index* and the other the *K index*. At each station, the lifted index is plotted above a short line and the K index below the line. An "M" indicates the value is missing.

The following explains the computation of the indices and the analysis and use of the panel. If you run into trouble with the discussion, you should review AVIATION WEATHER, chapter 6, "Stable and Unstable Air".

LIFTED INDEX (LI)

The lifted index is computed as if a parcel of air near the surface were lifted to 500 millibars. As the air is "lifted", it cools by expansion. The temperature the parcel would have at 500 millibars is then subtracted from the environmental 500 millibar temperature. The difference is the lifted index which may be positive, zero, or negative. Thus, the lifted index indicates stability at 500 mb. (18,000 feet MSL).

Positive Index

A positive index means that a parcel of a air *if lifted* would be colder than existing air at 500 millibars. The air is stable. Large positive (high) values indicate very stable air.

Zero Index

A zero index denotes that air *if lifted* to 500 millibars would attain the same temperature as the existing 500 millibar environmental temperature. The air is neutrally stable (neither stable nor unstable).

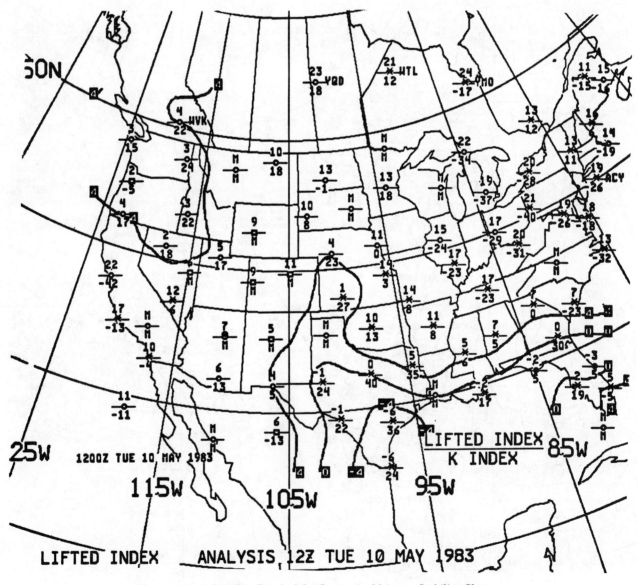

FIGURE 10-2. A Stability Panel of the Composite Moisture Stability Chart.

Negative Index

A negative index means that the low-level air *if lifted* to 500 millibars would be warmer than existing air at 500 millibars. The air is unstable and suggests the possibility of convection. Large negative (low) values indicate very unstable air.

Be aware of the following: 1) The LI assumes air near the surface will reach 500 millibars. Whether or not the near surface air will be lifted to 500 millibar level depends on what is happening below the level. It is possible to have a negative LI with no thunderstorms occurrence because the air below 500 millibars may not be lifted high enough for thunderstorms to develop. For use, the LI is more indicative of the severity of the thunderstorms, if they occur, rather than the probability of general thunderstorm occurrence (see TABLE 10-1). 2) The LI can change dramatically due to several causes, especially due to surface heating. Surface heating tends to make the LI value less positive (less stable) at 00Z as compared to 12Z.

K INDEX

The K index is primarily for the meteorologist but a discussion is included for those who are interested. The K index examines the temperature and moisture profile of the environment. Since a parcel of air is not lifted and compared to the enviroment, the K index is not truly a stability index. However, the meteorologist, looking at the environmental temperature and moisture profile, can make a good judgement as to the stability of the air. The K index is computed using three terms as follows:

$$K = (850\ mb\ temp - 500\ mb\ temp) + (850\ mb\ dew\ point) - (700\ mb\ temp\ dew\ point\ spread)$$

The first term (850mb temp—500 mb temp) is proportional to the average lapse rate. A large temperature difference shows a steep or unstable lapse rate. The greater the difference, the more unstable the air and the higher the K value.

The second term (850 mb dew point) is a measure of low level moisture. Since the dew point is added, high moisture content at 850 millibars increases the K value.

The third term (700 mb temp dew point spread) is a measure of saturation at 700 millibars. The greater the spread, the drier the air. Since the term is subtracted, it lowers the K value. However, moist air (small spread) lowers the value less than does dry air (large spread). Thus, the greater the degree of saturation at 700 millibars, the larger is the K value.

Putting the three terms together, we see that each of the following contributes to a large or high K index:
1. An unstable lapse rate
2. High moisture content at 850 millibars, and
3. A high degree of saturation at 700 millibars.

Since the K index is not a true stability index and with the moisture variables in the equation affecting the value and meaning of the index, great caution should be exercised as to when and how the K index is to be used. Thus, the K index is used primarily by the meteorologist. However, some general use may be made of the K index but only with caution.

During the thunderstorm season, a large K index indicates conditions favorable for air mass thunderstorms. See table 10-1 for thunderstorm potential. The K index values and meanings in table 10-1 can decrease significantly for thunderstorm development associated with a synoptic scale low pressure system (non air-mass thunderstorms).

In the winter, when temperatures are very cold, the moisture terms are very small. The temperature terms completely dominate the K value computation and even fairly large values do not mean conditions are favorable for thunderstorms because of lack of moisture.

Also be aware the K values can change significantly over a short time period due to temperature and moisture advection.

STABILITY ANALYSIS

The analysis is based on the lifted index only. Station circles are blackened for LI values of zero or

TABLE 10-1. Thunderstorm Potential

LIFTED INDEX (LI)	"K" INDEX *	AIRMASS THUNDERSTORM PROBABILITY
0 to −2 weak indication of severe thunderstorms	< 15	near 0%
	15-20	20%
−3 to −5 moderate indication of severe thunderstorms	21-25	21-40%
indication of severe thunderstorms	26-30	41-60%
≤−6 strong indication of severe thunderstorms	31-35	61-80%
	36-40	81-90%
	>40	near 100%

Note: See TABLE 4-9 for Areal Coverage Definitions

less. Solid lines are drawn for values of +4 and less at intervals of 4 (+4, 0, −4, −8 etc).

USING THE PANELS

As a user of stability indices, your question is "what helpful information can I obtain"? When clouds and precipitation are forecast or are occurring, the stability index is used to determine the type of clouds and precipitation. That is, stratiform clouds and steady precipitation occur with stable air while convective clouds and showery precipitation occur with unstable air. Remember that unstable air is associated with a negative lifted index and usually a high K index during the thunderstorm season (see table 10-1). Reliability of K index values decreases for non air-mass thunderstorms. Stability is also very important when considering the type, extent, and intensity of aviation weather hazards. For example, a quick estimate of areas of probable *convective* turbulence can be made by associating the areas with *unstable* air. An area of *extensive* icing would be associated with *stratiform* clouds and *steady* precipitation which are characterized by *stable* air.

It is essential to note that an unstable index does *not* automatically mean thunderstorms. Upon looking at the synoptic situation, if thunderstorms are expected to develop in unstable air, Table 10-1 may be used with caution as stated in this section. * Caution to be used for K index values in western mountainous terrain due to elevation.

FREEZING LEVEL PANEL

The freezing level panel (lower left panel of chart) is an analysis of observed freezing level data from upper air observations (see figure 10-3).

PLOTTED DATA

Table 10-2 explains plotting freezing level data. Note that more than one entry denotes multiple crossings of the zero (0) degree Celsius isotherm. See TABLE 10-3.

ANALYSIS

Solid lines are contours of the lowest freezing level and are drawn for 4000 foot intervals and labelled in hundreds of feet MSL. When a station reports more than one crossing of the zero (0) degree Celsius isotherm, the lowest crossing is used in the analysis. This is in contrast to the low level significant weather prog on which the depicted forecast freezing level aloft is the highest freezing level. A dashed line represents the 32 degree Fahrenheit isotherm

TABLE 10-2. Plotting freezing levels

Plotted	Interpretation
BF	Entire observation below freezing (0 degree C).
000	Surface temperature 0 degree C. Freezing level at surface and below freezing above.
120	Lowest and only freezing level 12,000 feet MSL; above freezing below 12,000 feet.
Three digits other than 000	Height of a freezing level aloft in hundreds of feet MSL, i.e.; 002, 200 feet MSL; 120, 12,000 feet MSL.
110 051 BF	Below freezing from surface to 5,100 feet; above freezing from 5,100 feet to 11,000 feet; and below freezing above 11,000 feet.
090 034 003	Lowest freezing level, 300 feet; below freezing from 300 feet to 3,400 feet; above freezing 3,400 feet to 9,000 feet; below freezing above 9,000 feet.
M	Data missing.
106 101 89 28 BF	Below freezing from surface to 2,800 ft.; above freezing from 2,800 ft. to 8,900 ft.; below freezing from 8,900 ft. to 10,100 ft.; above freezing from 10,100 ft. to 10,600 ft.; and below freezing above 10,600 ft.

at the surface and will outline an area of stations reporting "BF" (belowing freezing).

USING THE PANEL

The contour analysis shows an overall view of the lowest observed freezing level. Always plan for possible icing in clouds or precipitation above the freezing level—especially between temperatures of zero (0) Celsius and −10 Celsius.

Plotted multiple crossings of the zero (0) degree Celsius isotherm at a station always show an inversion with warm air above subfreezing temperatures (see TABLE 10-3). This situation can produce very hazardous icing when precipitation is occurring. Area forecasts show more specifically the areas of expected icing. Low level significant weather progs show anticipated changes in the freezing level.

PRECIPITABLE WATER PANEL

The precipitable water panel (upper right panel of chart) is an analysis of the water vapor content from

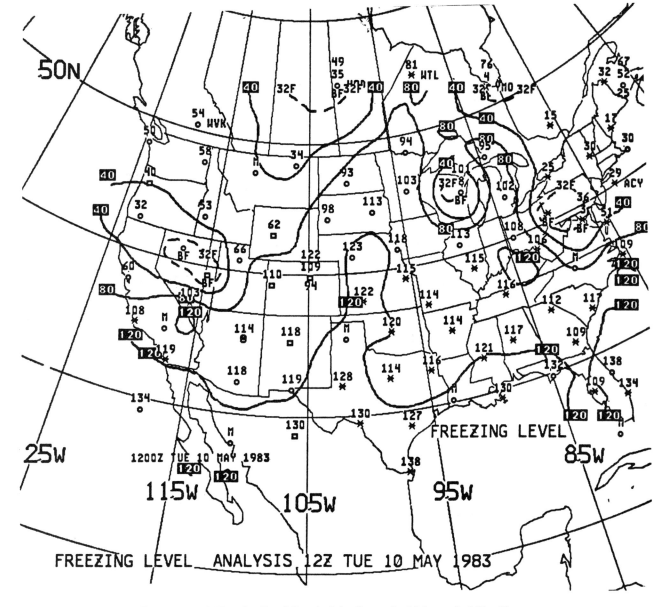

FIGURE 10-3. A Freezing Level Panel of the Composite Moisture Stability Chart.

the surface to the 500 mb level (see figure 10-4). The amount of water vapor observed is shown as precipitation water, which is the amount of liquid precipitation that would result if all the water vapor were condensed.

PLOTTED DATA

At each station precipitable water values to the nearest hundredth of an inch are plotted above a short line and the percent of normal value for the month below the line. the percent of normal value is the amount of precipitable water actually present compared to what is normally expected. As examples on figure 10-4, .58/69 at Oklahoma City OK indicates 58 hundredths of an inch of precipitable water is present which is only 69 percent of normal (below normal) for any day during this month, .81/1.45 at Dodge City KS indicates 81 hundredths of an inch of precipitable water is present which is 145 percent of normal (above normal) for any day during this month. An "M" plotted above the line indicates missing data as shown at Amarillo, TX. At Las Vegas NV the percent of normal value is not plotted which indicates insufficient climatological data to compute this value.

ANALYSIS

Stations with blackened in circles indicate precipitable water values of 1.00 inch or more. Isopleths of precipitable water are drawn and labelled for every

10-5

TABLE 10-3. Vertical temperature profile of plotted freezing levels at a station.

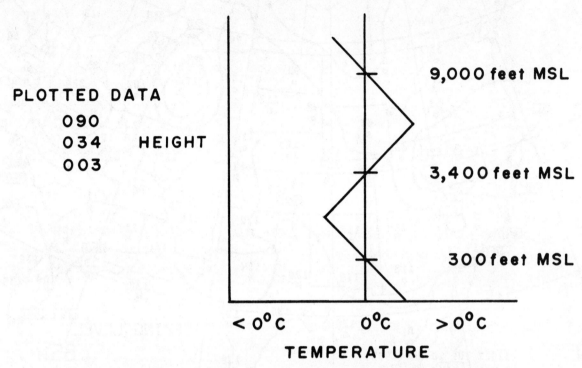

0.25 inches with the heavy isopleths drawn at 0.50 inch intervals.

USING THE PANEL

This panel is obviously used to determine water vapor content in the air between surface and 500 mb, and is especially useful to meteorologists concerned with flash flood events. By looking at the wind field upstream from your station, you can get an excellent indication of changes that will occur in moisture content, i.e. drying out or increasing moisture with time.

AVERAGE RELATIVE HUMIDITY PANEL

The average relative humidity panel (lower right panel of chart) is an analysis of the average relative humidity from the surface to 500 mb, plotted as a percentage for each reporting station (see figure 10-5). An "M" indicates the value is missing.

ANALYSIS

Station circles are blackened for humidities of 50 percent and higher. Isopleths of relative humidity (isohumes) are drawn and labelled every 10 percent with the heavy isohumes drawn for values of 10, 50 and 90 percent.

USING THE PANEL

This panel is used to determine, on the average, how saturated the air is from the surface to 500 mb. Average relative humidities of 70 percent or greater are frequently associated with areas of clouds and possible precipitation. This is because with such a

10-6

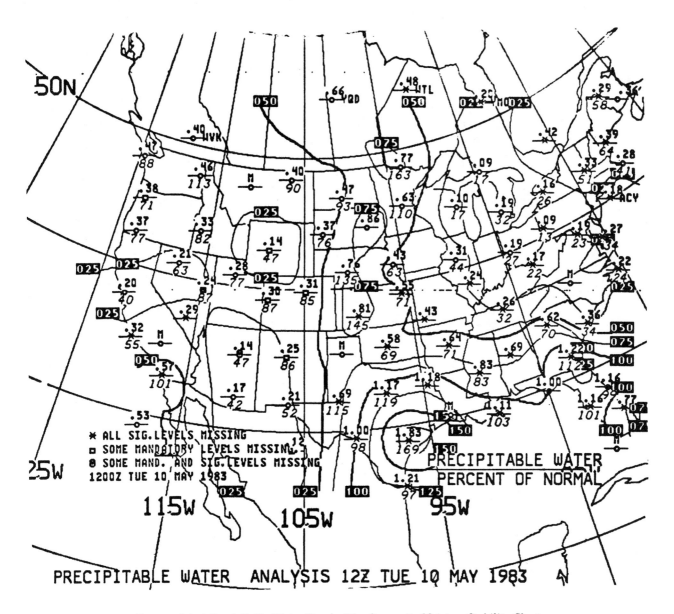

FIGURE 10-4. A Precipitable Water Panel of the Composite Moisture Stability Chart.

high average relative humidity through approximately 18,000 feet, it is likely that a specific layer(s) will have 100 percent relative humidity with clouds and possibly precipitation. It is important to remember that high values of relative humidity do not necessarily mean high values of water vapor content (precipitable water). For example in figure 10-4, the station in southwest Oregon has less water vapor content than International Falls MN (.37 and .77 respectively), but in examining figure 10-5, the average relative humidities are the same for both stations. If rain were falling at both stations, the result would likely be lighter precipitation totals for southwest Oregon.

USING THE COMPOSITE MOISTURE STABILITY CHART

Through analysis of this chart, you can determine the characteristics of a particular weather system in terms of stability, moisture, and possible aviation hazards. Even though this chart will be several hours old when received, the weather system will tend to move these characteristics with it. Thus, extrapolation techniques are an advantage to this chart although caution should be exercised due to modification of these characteristics through development, dissipation or the system moving from water to land or land to water.

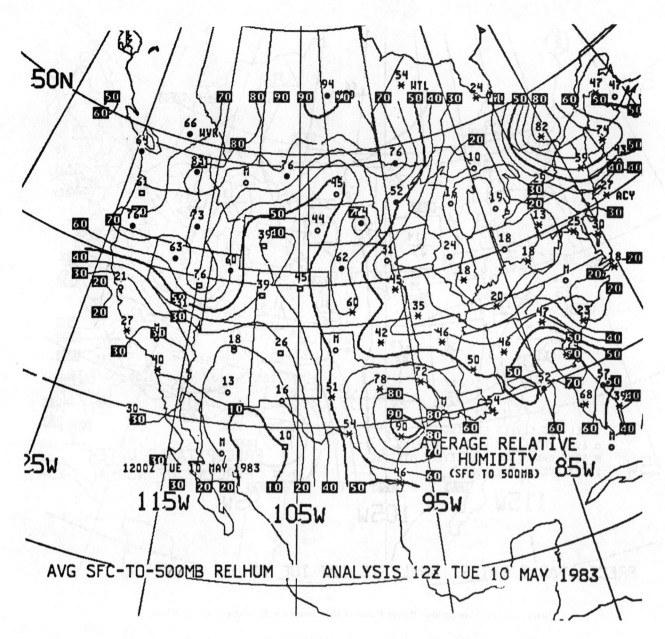

FIGURE 10-5. Average Relative Humidity Panel of the Composite Moisture Stability Chart.

Section 11
SEVERE WEATHER OUTLOOK CHART  (forecast)

The severe weather outlook chart, figure 11-1, is a preliminary 24-hour outlook for thunderstorm activity presented in two panels. The left-handed panel covers the 12-hour period 1200Z-0000Z. The right-hand panel covers the remaining 12 hours, 0000Z-1200Z. The manually prepared chart is issued once daily in the morning.

GENERAL THUNDERSTORMS

A line with an arrowhead delineates an area of probable general thunderstorm activity. When you face in the direction of the arrow, activity is expected to the right of the line. An area labelled APCHG indicates probable general thunderstorm activity may approach severe intensity. Approaching means winds greater than or equal to 35 knots but less than 50 knots and/or hail greater than or equal to 1/2 inch in diameter but less than 3/4 inch (surface conditions). Note in figure 11-1, from 12Z to 00Z that general thunderstorm activity is *not* forecast for the west coast, western Texas and from Tennessee northeastward to western New York.

SEVERE THUNDERSTORMS

The single-hatched area indicates possible severe thunderstorms. The following notations show possible coverage:

TABLE 11-1. Notation of Coverage

Notation	Coverage
SLIGHT RISK	2 to 5% coverage or 4 to 10 radar grid boxes containing severe thunderstorms per 100,000 square miles.
MODERATE RISK	6 to 10% coverage or 11 to 21 radar grid boxes containing severe thunderstorms per 100,000 square miles.
HIGH RISK	More than 10% coverage or more than 21 radar grid boxes containing severe thunderstorms per 100,000 square miles.

In figure 11-1, note the moderate risk of severe thunderstorms in the eastern Dakotas and Minnesota surrounded by a slight risk area. In the moderate risk area, severe thunderstorms are possible with severe storm coverage of 6 to 10 percent of the area.

TORNADOES

Tornado watches are plotted only if a tornado watch is in effect at chart time. The watch area is cross-hatched. No coverage is specified. Figure 11-1 shows a tornado watch in effect for eastern North Dakota and northern Minnesota at the time that the chart was issued.

USING THE CHART

The severe weather outlook is strictly for advanced planning. It alerts all interests to the possibility of future storm development. As the time of severe weather approaches, the forecaster can more specifically delineate the time, extent, and nature of the weather and issue a severe weather watch (WW).

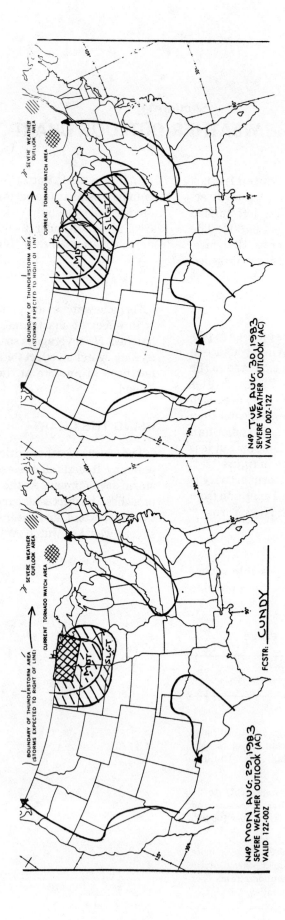

FIGURE 11-1. Severe Weather Outlook Charts.

Section 12
CONSTANT PRESSURE ANALYSIS CHARTS (observed)

Any surface of equal pressure in the atmosphere is a constant pressure surface. A constant pressure analysis chart is an upper air weather map where all the information depicted is at the specified pressure of the chart.

Twice daily, five (5) computer prepared constant pressure charts (850 mb, 700 mb, 500 mb, 300 mb and 200 mb) are transmitted by facsimile, each valid at 12Z and 00Z. Plotted at each reporting station (at the level of the specified pressure) are the observed temperature, temperature-dew point spread, wind, height of the pressure surface, as well as height changes over the previous 12 hour period. Figure 12-2 through 12-6 are sections of each constant pressure chart.

Pressure altitude (height in the standard atmosphere) for each of the five (5) pressure surfaces is shown in table 12-1. For example, 700 millibars of pressure has a pressure altitude of 10,000 feet (standard atmosphere). In the real atmosphere 700 millibars of pressure only closely approximates 10,000 feet (either above or below 10,000 feet) because the real atmosphere is seldom standard. For direct use of a constant pressure chart, assume you are planning a flight at 10,000 feet. The 700 mb chart is approximately 10,000 feet MSL. It is a source of observed temperature, temperature-dew point spread, moisture, and wind for your flight.

FIGURE 12-1. Radiosonde Data Station Plot and Decode.

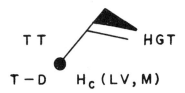

WIND—wind direction (WD) and speed (WS) plotted to the nearest 10 degrees and to the nearest 5 knots respectively.

⌐⎯₀ 5 knots, ⌐⎯⎯₀ 10 knots, ◣⎯₀ 50 knots

— if direction or speed is missing "M" is plotted in H_c space.
— if speed is less than 3 knots "LV" (light and variable) is plotted in H_c space.

HGT —height of constant pressure surface in meters. See TABLE 12-1 for decoding. If data is missing nothing is plotted in this spot.

TT —temperature to the nearest whole degree C; minus sign used if negative. Left blank if TT is missing. On the 850 mb chart, primarily in mountain regions where stations may be located above 850 mb of pressure, a bracketed temperature (and HGT) is a computed value. If two temperatures are plotted one above the other, the top temperature is used in the analysis.

T-D —temperature-dew point spread (or depression) to the nearest whole degree C. Left blank if T-D is missing. If T-D is less than or equal to 5 degrees C, the station circle is completely blackened. If T-D is greater than 29 degrees C and "X" is plotted. If TT is colder than −41 degrees C, T-D is left blank because the air is too dry at those temperatures to measure dew point.

H_c —Previous 12 hour height change plotted in tens of meters (decameters). H_c not plotted when wind is "LV" or "M". +04 means height of pressure above station has risen 40 meters, 02=meters, 11=110 meters.

LV = Light & variable
M = missing

PLOTTED DATA

Figure 12-1 illustrates and decodes the standard radiosonde data plot. Table 12-2 gives examples. Aircraft and satellite observations are used in analysis over areas of sparse data. A square in lieu of a station circle signifies an aircraft report. The flight level of the aircraft is plotted in hundreds of feet. Temperature and wind are at the flight level of the aircraft. The time of report is also indicated to the nearest hour GMT. For example, in figure 12-5, the aircraft report at approximately 30 degrees N and 140 degrees W is decoded as follows: flight level 34,000 feet, temperature −50 degrees Celsius, wind from 280 degrees at 10 knots, time of report to nearest hour is 2100 GMT. A star identifies satellite wind estimates made from cloud tops and is used primarily on the 850 mb chart off the west coast. These winds are representative for the 850 mb chart even though they are always labelled with a pressure altitude of 3,000 feet. See figure 12-2 for examples.

12-1

TABLE 12-1. Features of constant pressure charts-U.S.

PRESSURE (millibars)	PRESSURE ALTITUDE in feet (flight level)	PRESSURE ALTITUDE In meters	TEMPERATURE DEW POINT SPREAD	ISOTACHS	CONTOUR INTERVAL (meters)	DECODE STATION HEIGHT PLOT		EXAMPLES OF STATION HEIGHT PLOTTING	
						PREFIX TO PLOTTED VALUE	SUFFIX TO PLOTTED VALUE	PLOTTED	HEIGHT
850	5,000	1,500	YES	NO	30	1	—	530	1,530
700	10,000	3,000	YES	NO	30	2 or 3*	—	180	3,180
500	18,000	5,500	YES	NO	60	—	0	582	5,820
300	30,000	9,000	YES**	YES	120	—	0	948	9,480
200	39,000	12,000	YES**	YES	120	1	0	164	11,640

Note that pressure altitudes are rounded off to the nearest thousand for feet and to the nearest 500 for meters.

All heights are MSL.

* Prefix a "2" or "3" whichever brings the height closer to 3,000 meters.
** Omitted when air is too cold (less than -41 degrees) to measure dew point. Flight level of an aircraft is plotted in lieu of height of constant pressure surface.

TABLE 12-2. Examples of radiosonde plotted data.

	(850 MB)	(700 MB)	(500 MB)	(300 MB)	(200 MB)
WIND	LIGHT AND VARIABLE	010° 20KTS	210° 60KTS	270° 25KTS	MISSING
TT	22°C	9°C	-19°C	-46°C	-60°C
T-D	4°C	17°C	) 29°C	not plotted	not plotted
DEW POINT	18°C	-8°C	DRY	DRY	DRY
HGT	1,479 meters	3,129 meters	5,580 meters	9,190 meters	11,910 meters
H$_c$	not plotted	MINUS 30 meters	PLUS 30 meters	+100 meters	not plotted

12-2

ANALYSIS

All charts contain contours and isotherms and some contain isotachs. Contours are lines of equal height, isotherms are lines of equal temperature, and isotachs are lines of equal wind speed.

Height Contours

Heights of the specified pressure for each station are analyzed through the use of solid lines called contours to give a *height* pattern. The contours depict highs, lows, troughs and ridges aloft in the same manner as isobars on the surface chart. Thus, on an upper air chart we speak of "high height centers" and "low height centers" instead of "high pressure centers" and "low pressure centers" respectively. We may compare a height analysis to a pressure analysis. A *contour* high, low, ridge or trough and the two terms are used interchangebly. For example, a high height center at 500 mb of pressure is analogous to a high pressure center at about 18,000 feet. Height and pressure analyses are just two ways of describing the same features.

Since an upper air chart is above the surface friction layer, winds for practical purposes parallel the contours. To decode contour values on the 850 mb through 300 mb chart simply add a zero to the three digit code while on the 200 mb chart you must prefix a one (1) in addition to adding a zero.

Refer to figures 12-2 and 12-3 and note the low aloft in Montana extending upward through 700 mb. Figures 12-4 through 12-6 show the low has opened into a trough at 500 mbs and above. Note that this low tilts toward the west with height.

Isotherms

Isotherms (dashed lines) drawn at 5 degrees Celsius intervals show horizontal temperature variations at chart altitude. Let's refer to the 850 mb chart (5,000 feet pressure altitude), figure 12-2 and locate an isotherm. Note the dashed line extending from North Dakota southeastward through South Carolina and labelled "+10" in the western portion of North Dakota. This is the +10 degree Celsius isotherm. North of this isotherm, temperatures at approximately 5,000 feet are below +10 degrees Celsius. The +15 degree Celsius isotherm extends across South Dakota. By inspecting isotherms, you can determine if your flight will be toward colder or warmer air. Subfreezing temperatures and a temperature-dew point spread of 5 degrees Celsius or less suggest possible icing. Note that isotherms on the 300 and 200 mb charts are the heavy dashed lines.

Isotachs

Isotachs (short, lightly dashed lines) appear only on the 300 and 200 mb charts. Isotachs are drawn at 20 knot intervals beginning with 10 knots. To aid in identifying areas of strong winds, hatching denotes wind speeds of 70 to 110 knots, a clear area within a hatched area indicates wind or 110 to 150 knots, an area of 150 to 190 knots of wind is hatched, etc. Note the alternating hatched/clear areas in figure 12-7 extending eastward from California to Florida and up the east coast. The 150 knots isotach (150 K) is over southern New Mexico and across Oklahoma and Texas.

DO NOT confuse isotherms with isotachs.

THREE DIMENSIONAL ASPECTS

As established earlier, we may treat a height contour analysis as a pressure analysis. Closely spaced contours mean strong winds as do closely spaced isobars. Wind blows clockwise around a contour high and counterclockwise around a low.

Features on synoptic surface and upper air charts are related. However, a weak surface system often loses its identity in a large scale upper air pattern while another system may be more evident on an upper chart than on the surface chart. Many times weather is closely associated with an upper air pattern than with features on the surface map.

You may have learned as a general rule to regard a surface low as a producer of bad weather and a high as a producer of good weather. Usually, this is true but an upper level low or trough usually means bad weather also. The area of cloudiness and precipitation found with an upper air low (on the east side) is usually associated with a surface low but sometimes an upper level low with clouds and precipitation will move over a rather shallow surface high with corresponding bad weather in the high. In contrast, an upper air high usually means good weather. An exception is an upper air high or ridge that has a stabilizing effect at low levels. Smoke, haze, dust, or even low stratus and fog may persist for extended periods; yet the surface map shows no cause for the restriction.

Lows generally slope to the west with ascending altitude for developing low pressure systems. Due to this slope, wind aloft with an upper system often blows across the associated surface system. Surface fronts, lows, and highs tend to move with the upper winds. For example, strong winds aloft across a surface front will cause the front to move rapidly, but if upper winds parallel a front, it moves slowly if at all.

An old, non-developing low pressure system tilts little with height. The low becomes almost vertical and is clearly evident on both surface and upper air maps. Upper winds encircle the surface low rather than blow across it. Thus, the storm moves very slowly and usually causes extensive and persistent cloudiness, precipitation, and generally adverse flying weather. The term "cold low" describes such a system and is usually identified on the surface chart

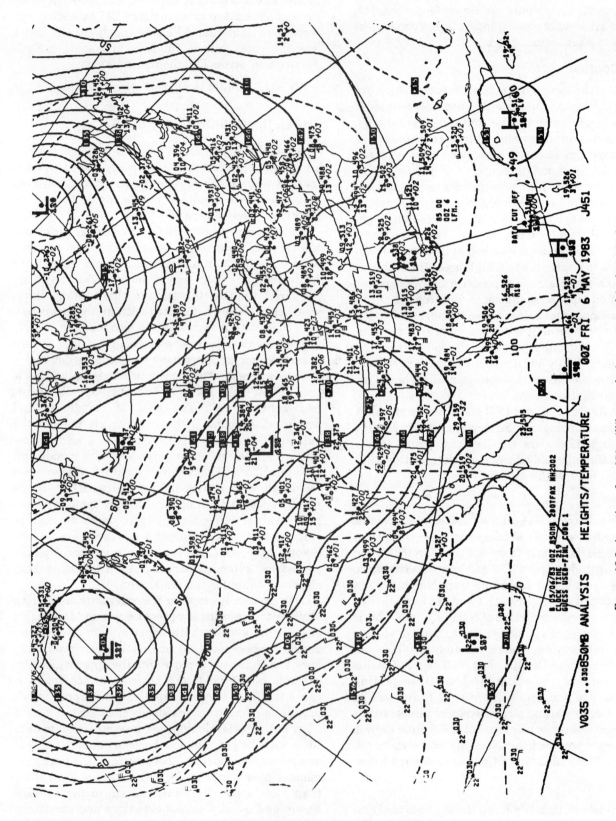

FIGURE 12-2. A section of an 850 millibar analysis, pressure altitude 5,000 feet.

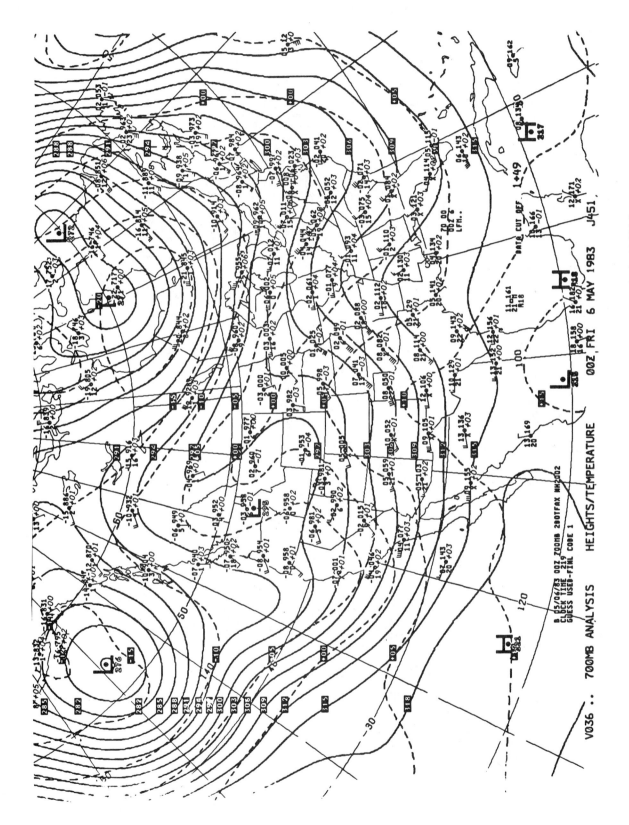

FIGURE 12-3. A section of a 700 millibar analysis, pressure altitude 10,000 feet.

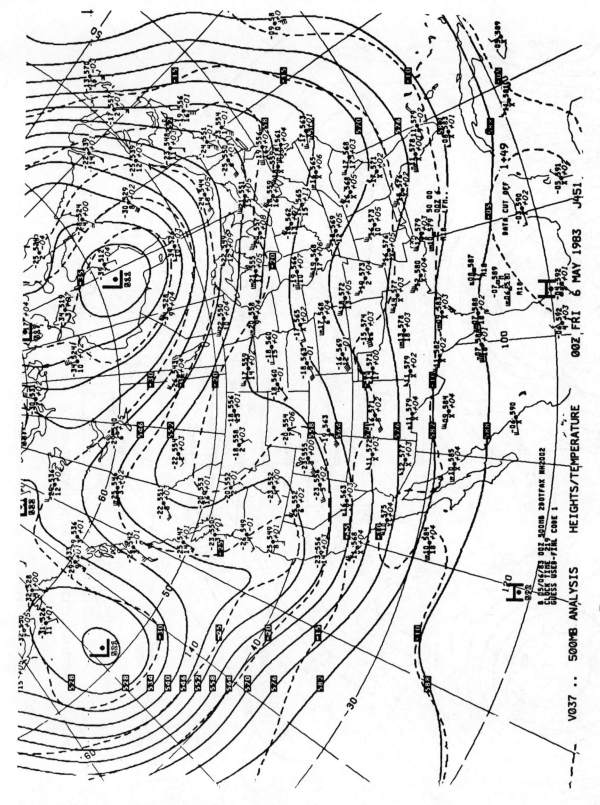

FIGURE 12-4. A section of a 500 millibar analysis, pressure altitude 18,000 feet.

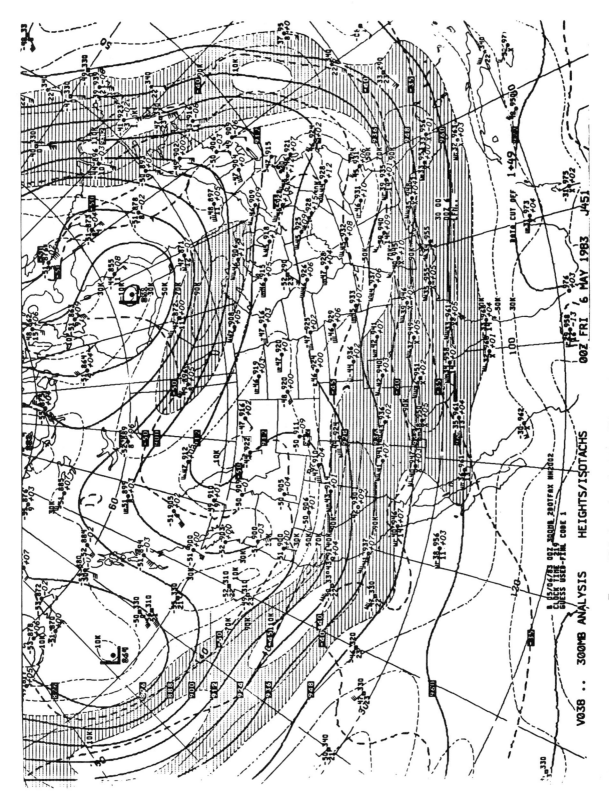

FIGURE 12-5. A section of a 300 millibar analysis, pressure altitude 30,000 feet.

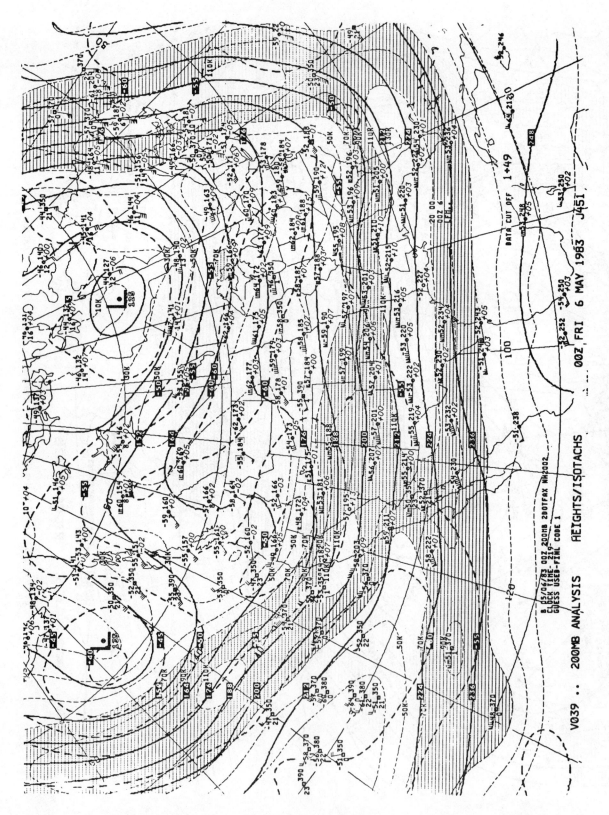

Figure A section of a 200 millibar analysis, pressure altitude 39,000 feet.

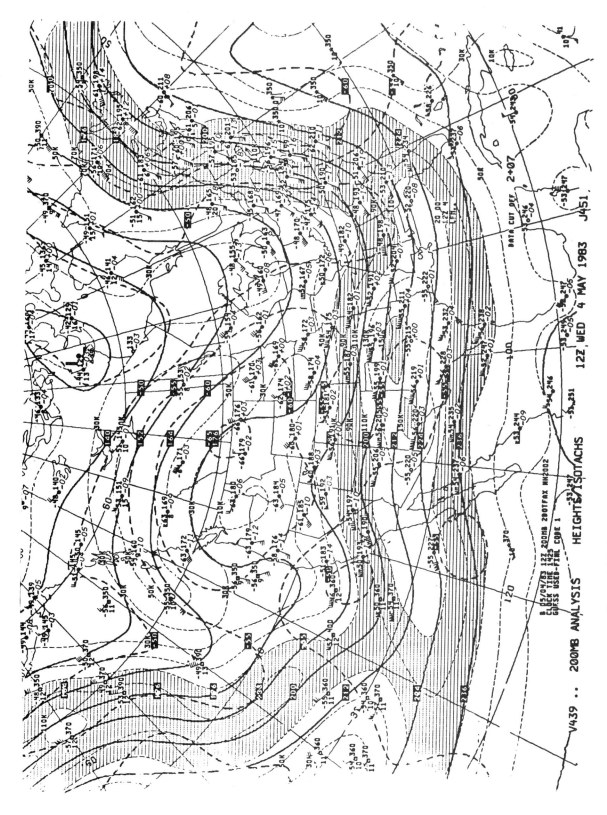

FIGURE 12.7. A section of a 200 millibar analysis, pressure altitude 39,000 feet.

as an old, occluded low with the warm air having been cut-off from the low pressure center.

In contrast to the cold low is the "thermal low". A dry, sunny region becomes quite warm from intense surface heating resulting in a surface low pressure area. The warm air is carried to high levels by convection but cloudiness is scant because of lack of moisture. The warm surface low often is "capped" by a high aloft. Unlike the cold low, the thermal low is relatively shallow with weak pressure gradients and no well defined cyclonic circulation. However, you must be alert for high density altitude, light to moderate convective turbulence, and isolated showers and thunderstorms if sufficient moisture is present. The thermal low is a semipermanent feature of the desert regions in the southwestern United States and northern Mexico during warm weather.

These are only a few rough examples of associating weather with upper air features. They point out the need to view weather in three dimensions; to get a "picture" of the atmosphere which is the first step to understanding the atmosphere and its weather.

USING THE CHARTS

From the charts you can approximate the observed temperature, wind, and temperature-dew point spread along your proposed route. Usually you can select a constant pressure chart close to your planned altitude. For altitudes about midway between two charted surfaces, interpolate between the two charts.

Determine temperature from plotted data or the pattern of isotherms. To readily delineate areas of high moisture content, station circles are shaded indicating temperature-dew point spreads of 5 degrees Celsius or less. You can get the actual spread from plotted data. A small spread alerts you to possible cloudiness, precipitation and icing. Determine windspeed for lower levels from plotted data, for the 300 and 200 millibar surfaces, determine wind speed from the isotach pattern. Wind direction parallels the contours. Using isotherms and contours, you can determine thermal advection (warming/cooling with time).

As stated earlier, constant pressure charts often show the cause of weather and its movement more clearly than does the surface map. For example, the large scale wind flow around a low aloft may spread cloudiness, low ceilings, and precipitation far more extensively than indicated by the surface map alone.

Keep in mind that constant pressure charts are observed weather.

Section 13
TROPOPAUSE DATA CHART  (observed)

A four (4) panel chart containing observed tropopause data, a maximum wind prog, a vertical wind shear prog, and a high level significant weather prog is prepared for the contiguous 48 states, see figure 13-1. The high level significant weather prog was covered in Section 8. The chart is available twice daily with observed data valid at 00Z and 12Z. Progs are 18 hour forecasts valid at 18Z to 06Z.

OBSERVED TROPOPAUSE PANEL

The observed tropopause data panel shows for each upper air observing station the pressure, temperature, and wind at the tropopause. Figure 13-2 shows the panel with Alburquerque NM (ABQ) identified to aid in explaining the station model. Decode the plotted data at Alburquerque as follows: tropopause wind, 240 degrees, at 115 knots; tropopause temperature, −61 degrees Celsius; tropopause pressure, 200 millibars.

USING THE PANEL

Maximum wind occurs near the tropopause, so this panel is essentially a map of observed maximum winds. A close inspection of the map reveals a jet stream from central New Mexico across southeastern Kansas, central Missouri, and southern Illinois to West Virginia. The reason wind data are missing over Oklahoma, eastern Kansas, and Illinois is that strong winds carried the radiosonde instruments too far from observing stations to obtain reliable wind data. This area of missing wind data is actually the area of strongest winds in the jet stream.

From the map you can determine observed wind and temperature at the tropopause. You can then use constant pressure progs or the FD winds and temperatures aloft forecast to interpolate for a flight level between a constant pressure level and the tropopause.

DOMESTIC TROPOPAUSE WIND AND WIND SHEAR PROGS

Forecast parameters at the tropopause over the contiguous 48 states and some adjoining oceanic and North American areas are shown on two panels—the *tropopause winds* and the *tropopause height/vertical wind shear progs*.

TROPOPAUSE WINDS

The tropopause winds prog, figure 13-3, depicts wind direction by streamlines—solid lines. Streamlines have no dimensions and are unlabelled. They are of sufficient density to show the direction field. Winds parallel the streamlines. Direction of the streamlines basically is from west to east in mid latitudes. A high or low may be encircled by a closed streamline; you can readily determine whether it is a high or low if you know the circulation around these systems.

Wind speed is shown by isotachs at 20 knots intervals (dashed lines in figure 13-3). They are labelled in knots. Areas of wind speeds between 70 and 110 knots are hatched as are wind speeds between 150 and 190 knots. The shading criteria are the same as used on selected constant pressure analysis and progs.

TROPOPAUSE HEIGHT/VERTICAL WIND SHEAR

Tropopause height/vertical wind shear prog (figure 13-4) depicts height of the tropopause in terms of pressure altitude and vertical wind shear in knots per 1,000 feet, (see chapter 3, AVIATION WEATHER, discussion of pressure altitude). Solid lines trace intersections of the tropopause with standard constant pressure surfaces. Heights are preceded by the letter F and are in hundreds of feet.

Following is a listing of pressure and corresponding flight levels:

Millibars	Flight Level
500	18,000
450	21,000
400	24,000
350	27,000
300	30,000
250	34,000
200	39,000
150	45,000
100	53,000
70	63,000

Vertical wind shear is in knots per 1,000 feet depicted by dashed lines at 2 knot intervals. Wind shear is averaged through a layer from about 8,000 feet below to 4,000 feet above the tropopause.

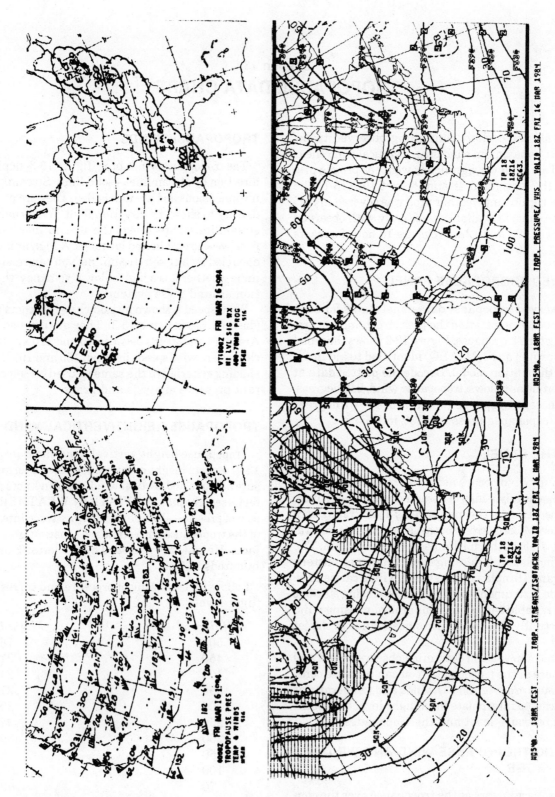

FIGURE 13-1. Tropopause data chart. Panels of observed tropopause data, maximum wind prog, vertical wind shear prog, and high level significant weather prog.

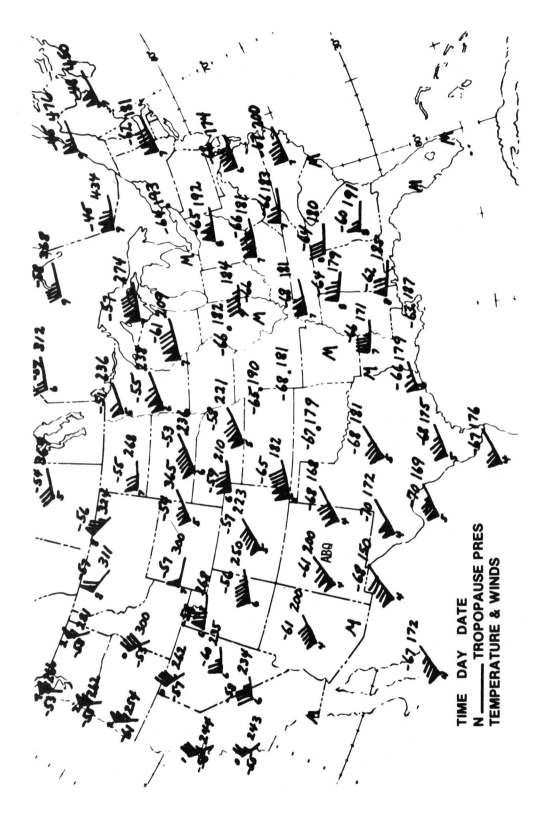

FIGURE 13-2. An observed tropopause panel.

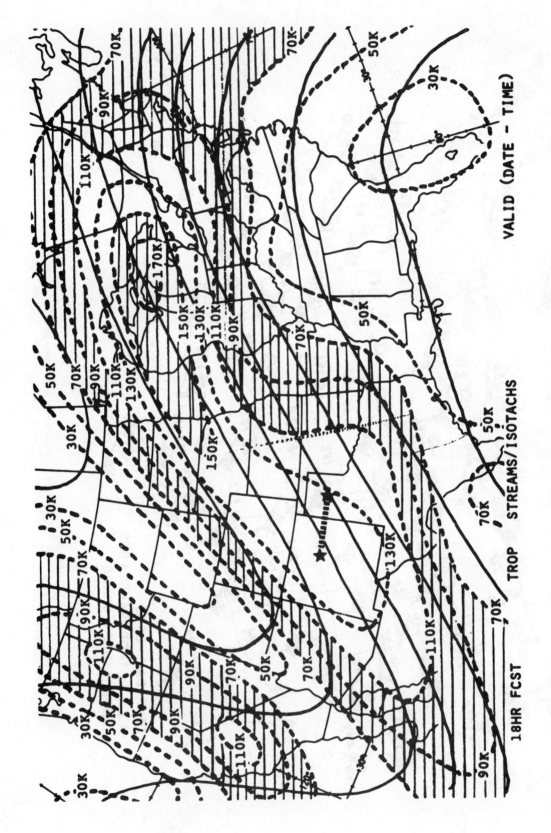

FIGURE 13-3. Section of a tropopause wind prog.

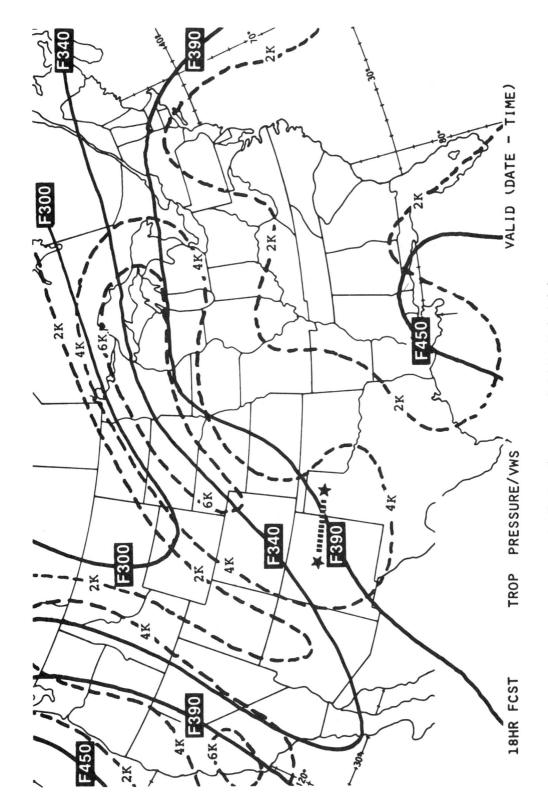

FIGURE 13-4. Section of a tropopause height/vertical wind shear prog.

13-5

USING THE PANELS

The progs are issued twice daily and may be used for a period up to plus or minus 6 hours from the valid time. The panels may be used to determine vertical and horizontal wind shears as clues to probable wind shear turbulence (see pages 14-4 and 14-5 for criteria). They also may be used to determine winds for high level flight planning.

Although neither panel depicts the jet stream, locating the jet is not difficult. It passes through isotach and vertical shear maxima. Examine figures 13-3 and 13-4, note a jet maximum from eastern Washington and Oregon extending southward and slightly westward through central California. It reappears near the southwest corner of the panels, enters the U.S. near the Arizona-New Mexico border, extends northeastward across central Nebraska, and then swings more easterly through the central Great Lakes and southern New England.

Horizontal wind shear can be determined from the spacing of isotachs. The horizontal wind shear critical for turbulence (moderate or greater) is greater than 18 knots per 150 miles (see chapter 13, AVIATION WEATHER, discussion on clear air turbulence). 150 nautical miles is equal to 2 1/2 degrees latitude.

Refer to figure 13-3 and measure 2 1/2 degrees latitude by laying a pencil along a meridian in the Atlantic. Move the pencil perpendicular to the isotach across north central Montana and you can see that the horizontal shear (the difference in wind speed) is about 40 knots along this distance. This spacing represents the wind shear critical for probable moderate or greater wind shear turbulence. The strong wind shear from southwestern Arizona to northwestern Minnesota suggests a probability of turbulence due to horizontal wind shear.

Vertical wind shear can be determined directly from the dashed lines in figure 13-4. The vertical shear critical for probable turbulence is 6 knots per 1,000 feet. You find this critical value in central California and from western Nebraska to the Great Lakes. An area of extremely high probability of moderate or greater turbulence is the three state junction of the Dakotas and Minnesota where horizontal shear is about 80 knots per 150 miles and vertical shear is in excess of 6 knots per 1,000 feet.

Wind direction and speed at the tropopause flight level may be read directly from the streamlines and isotachs. To determine wind at a flight level below and above the tropopause, first determine direction and speed at the tropopause. Wind direction changes very little within several thousand feet of the tropopause, so this direction may be used throughout the layer for which vertical wind shear is computed. Next determine wind shear and the number of thousands of feet the desired flight level differs from flight level of the tropopause. Multiply the shear by thousands of feet and subtract this value from the speed at the tropopause.

As an example, let's assume a westbound flight wants the probability of turbulence and the wind for a leg from Amarillo to Alburquerque. Note from the panels in figures 13-3 and 13-4 that horizontal wind shear is negligible. Vertical wind shear is interpolated between the 4 and 6 knot shear lines and is about 5 knots per 1,000 feet. Widespread significant turbulence (moderate or greater) is unlikely. You should also refer to the high level significant weather prog and pilot reports for further clues to turbulence.

Wind direction along the route determined from the streamline is about 230 degrees, a quartering headwind. Speed is strongest at the tropopause, so for a westbound flight, choose a flight level as far as practical above or below the tropopause. Height of the tropopause determined from figure 13-2 is flight level 39,000 feet (200 millibars). Let's assume you would like to check the wind at 43,000 feet. From figure 13-3, you determine tropopause wind speed to be on the high side of the 130 knot isotach but quite a distance from the 150 knot isotach. Let's interpolate the speed as 135 knots. The flight level, 43,000 feet, is 4,000 feet above the tropopause. Multiply the 5 knot shear by 4 and you get a difference of 20 knots. Subtract 20 knots from 135, the speed at the tropopause, and you get a speed of 115 knots. Therefore, wind at FL430 is 230 degrees at 115 knots.

Section 14
TABLES AND CONVERSION GRAPHS

This section provides graphs and tables you can use operationally in decoding weather messages during preflight and inflight planning and in transmitting pilot reports. Information included covers:
1. Icing intensities and reporting.
2. Turbulence intensities and reporting.
3. Locations of probable turbulence by intensity versus weather and terrain features.
4. Standard temperature, speed, and pressure conversions.
5. Density altitude computations.
6. Selected contractions.
7. Selected acronyms.
8. Scheduled issuance and valid times of forecast products.

The table of *Icing Intensities* classifies each intensity according to its operational effects on aircraft.

The table of *Turbulence Intensities* classifies each intensity according to its effects on aircraft control and structual integrity and on articles and occupants within the aircraft.

The table of *Locations of Probable Turbulence* lists each turbulence intensity along with terrain and weather features conducive to turbulence of that intensity.

The graph for *Density Altitudes Computations* provides a means of computing density altitude, either on the ground or aloft, using the aircraft altimeter and outside air temperature.

Contractions are used extensively in surface, radar, and pilot reports and in forecasts. Most of them are known from common usage or can be deciphered phonetically. The list of *Selected Contractions* contains only those most likely to give you difficulty. Acronyms used in this manual are defined in the list of *Acronyms*.

Locations of Probable Turbulence by Intensities Versus Weather and Terrain Features

LIGHT TURBULENCE
1. In hilly and mountainous areas even with light winds.
2. In and near small cumulus clouds.
3. In clear-air convective currents over heated surfaces.

TABLE 14-1. Icing intensities, airframe ice accumulation and pilot report

Intensity	Airframe ice accumulation	Pilot report
Trace	Ice becomes perceptible. Rate of accumulation slightly greater than rate of sublimation. It is not hazardous even though deicing/anti-icing equipment is not used unless encountered for an extended period of time—over one hour.	Aircraft identification, location, (GMT), intensity and type of icing *, altitude/FL, aircraft type, IAS
Light	The rate of accumulation may create a problem if flight is prolonged in this environment (over one hour). Occasional use of deicing/anti-icing equipment removes/prevents accumulation. It does not present a problem if the deicing/anti-icing equipment is used.	
Moderate	The rate of accumulation is such that even short encounters become potentially hazardous and use of deicing/anti-icing equipment or diversion is necessary.	Example of pilot's transmission: Holding at Westminister VOR 1232. Light Rime Icing. Altitude six thousand, Jetstar IAS 200 kt
Severe	The rate of accumulation is such that deicing/anti-icing equipment fails to reduce or control the hazard. Immediate diversion is necessary.	

* Icing may be rime, clear or mixed.
Rime ice: Rough milky opaque ice formed by the instanteous freezing of small super cooled water droplets.
Clear ice: A glossy, clear or translucent ice formed by the relatively slow freezing of large supercooled water droplets.
Mixed ice: A combination of rime and clear ice.

14-1

TABLE 14-2. Turbulence reporting criteria

Intensity	Aircraft reaction	Reaction inside aircraft	Reporting term-definition
Light	Turbulence that momentarily causes slight, erratic changes in altitude and/or attitude (pitch, roll, yaw). Report as *Light Turbulence;** o r Turbulence that causes slight, rapid and somewhat rhythmic bumpiness without appreciable changes in altitude or attitude. Report as *Light Chop*.	Occupants may feel a slight strain against seat belts or shoulder straps. Unsecured objects may be displaced slightly. Food service may be conducted and little or no difficulty is encountered in walking.	Occasional—Less than 1/2 of the time. Intermittent—1/3 to 2/3. Continous—More than 2/3.
Moderate	Turbulence that is similar to Light Turbulence but of greater intensity. Changes in altitude and/or attitude occur but the aircraft remains in positive control at all times. It usually causes variations in indicated airspeed. Report as *Moderate Turbulence;** or Turbulence that is similar to Light Chop but of greater intensity. It causes rapid bumps or jolts without appreciable changes in aircraft altitude or attitude. Report as *Moderate Chop*.	Occupants feel definite strains against seat belts or shoulder straps. Unsecured objects are dislodged. Food service and walking are difficult.	NOTE 1. Pilots should report location(s), time (GMT), intensity, whether in or near clouds, altitude, type of aircraft and, when applicable, duration of turbulence. 2. Duration may be based on time between two locations or over a single location. All locations should be readily identifiable.
Severe	Turbulence that causes large, abrupt changes in altitude and/or attitude. It usually causes large variations in indicated airspeed. Aircraft may be momentarily out of control. Report as *Severe Turbulence.**	Occupants are forced violently against seat belts or shoulder straps. Unsecured objects are tossed about. Food service and walking are impossible.	EXAMPLES: a. Over Omaha, 1232Z, Moderate Turbulence, in cloud, Flight Level 310, B707. b. From 50 miles south of Alburquerque to 30 miles north of Phoenix, 1210Z to 1250Z, occasional Moderate Chop, Flight Level 330, DC8.
Extreme	Turbulence in which the aircraft is violently tossed about and is practically impossible to control. It may cause structural damage. Report as *Extreme Turbulence.**		

* High level turbulence (normally above 15,000 feet AGL) not associated with cumuliform cloudiness, including thunderstorms, should be reported as CAT (clear air turbulence) preceded by the appropriate intensity, or light or moderate chop.

STANDARD CONVERSION

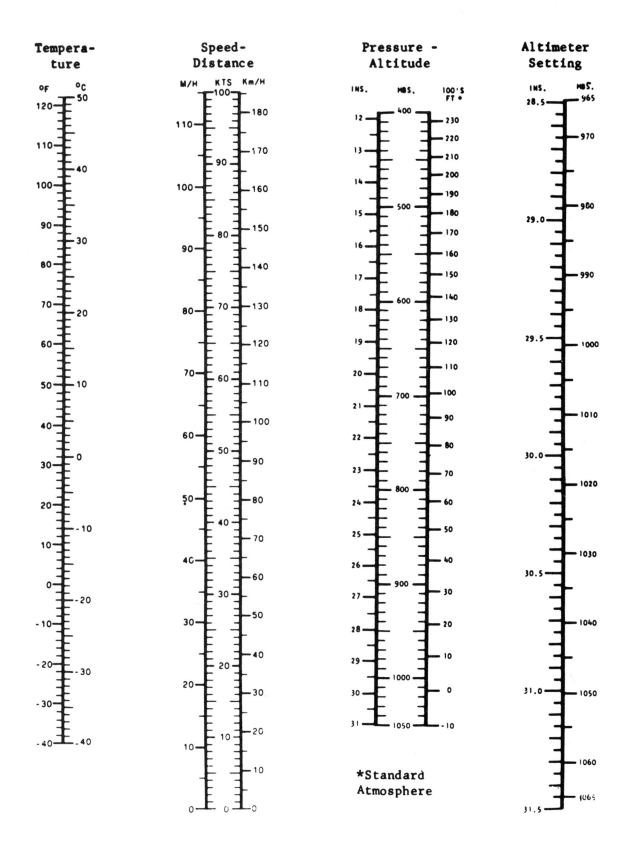

*Standard Atmosphere

4. With weak wind shears in the vicinity of:
   a. Troughs aloft.
   b. Lows aloft.
   c. Jet streams.
   d. The tropopause.
5. In the lower 5,000 feet of the atmosphere:
   a. When winds are near 15 knots.
   b. Where the air is colder than the underlying surfaces.

MODERATE TURBULENCE

1. In mountainous areas with a wind component of 25 to 50 knots perpendicular to and near the level of the ridge:
   a. At all levels from the surface to 5,000 feet above the tropopause with preference for altitudes:
      (1) Within 5,000 feet of the ridge level.
      (2) At the base of relatively stable layers below the base of the tropopause.
      (3) Within the tropopause layer.
   b. Extending outward on the lee of the ridge for 150 to 300 miles.
2. In and near thunderstorms in the dissipating stage.
3. In and near other towering cumuliform clouds.
4. In the lower 5,000 feet of the tropopause:
   a. When surface winds are 30 knots or more.
   b. Where heating of the underlying surface is unusually strong.
   c. Where there is an invasion of very cold air.
5. In fronts aloft.
6. Where:
   a. Vertical wind shears exceed 6 knots per 1,000 feet, and/or
   b. Horizontal wind shears exceed 18 knots per 150 miles.

SEVERE TURBULENCE

1. In mountainous areas with a wind component exceeding 50 knots perpendicular to and near the level of the ridge:
   a. In 5,000—foot layers:
      (1) At and below the ridge level in rotor clouds or rotor action.
      (2) At the tropopause.
      (3) Sometimes at the base of other stable layers below the tropopause.
   b. Extending outward on the lee of the ridge for 50 to 150 miles.
2. In and near growing and mature thunderstorms.
3. Occasionally in other towering cumuliform clouds.
4. 50 to 100 miles on the cold side of the center of the jet stream, in troughs aloft, and in lows aloft where:
   a. Vertical wind shears exceed 6 knots per 1,000 feet, and
   b. Horizontal winds shears exceed 40 knots per 150 miles.

EXTREME TURBULENCE

1. In mountain wave situations, in and below the level of well-developed rotor clouds. Sometimes it extends to the ground.
2. In severe thunderstorms (most frequently in organized squall lines) indicated by:
   a. Large hailstones (3/4 inch or more in diameter).
   b. Strong radar echoes, or
   c. Almost continuous lightning.

DENSITY ALTITUDE COMPUTATION

Use this graph to find density either on the ground or aloft. Set your altimeter at 29.92 inches; it now indicates pressure altitude. Read outside air temperature. Enter the graph at your pressure altitude and move horizontally to the temperature. Read density altitude from the sloping lines.

Example 1.

Find density altitude in flight. Pressure altitude is 9,500 feet; and temperature, −8 degrees C. Find 9,500 feet on the left of the graph and move across to −8 degrees C. Density altitude is 9,000 feet (marked "1" on the graph).

Example 2.

Find density altitude for take-off. Pressure altitude is 4,950 feet; and temperature 97 degrees F. Enter the graph at 4,950 feet and move across to 97 degrees F. Density altitude is 8,200 feet (marked "2" on graph). Note that in the warm air, density altitude is considerably higher than pressure altitude.

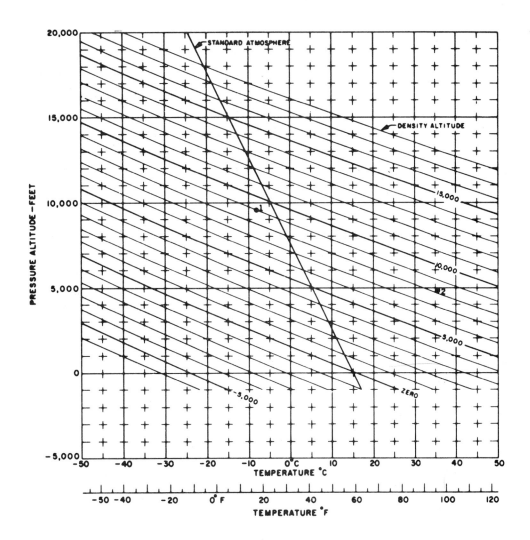

SELECTED CONTRACTIONS

A

ACLD	above clouds
ACSL	standing lenticular altocumulus
ACYC	anticyclonic
AFDK	after dark
ALQDS	all quadrants
AC	altocumulus
ACCAS	altocumulus castellanus
AS	altostratus
AOA	at or above
AOB	at or below

B

BCKG	backing
BFDK	before dark
BINOVC	breaks in overcast
BL	between layers
BLZD	blizzard
BOVC	base of overcast

C

CBMAM	cumulonimbus mamma
CC	cirrocumulus
CCSL	standing lenticular cirrocumulus
CFP	cold frontal passage
CI	cirrus
CLRS	clear and smooth
CRLCN	circulation
CS	cirrostratus
CU	cumulus
CUFRA	cumulus fractus
CYC	cyclonic

D

DFUS	diffuse
DNSLP	downslope
DP	deep
DTRT	deteriorate
DURGC	during climb

14-5

DURGD	during descent	**O**	
DWNDFTS	downdraft	OAOI	on and off instruments
		OAT	outside air temperature
E		OCFNT	occluded front
EMBDD	embedded	OCLD	occlude
		OFP	occluded frontal passage
F		OFSHR	off shore
FNTGNS	frontogenesis (front forming)	OI	on instruments
FNTLYS	frontolysis (front decaying)	OMTNS	over mountains
FROPA	frontal passage	ONSHR	on shore
		OTAS	on top and smooth
G		OVRNG	overrunning
GFDEP	ground fog estimated—feet deep		
		P	
H		PDW	priority delayed weather
HDEP	haze layer estimated—feet deep	PRESFR	pressure falling rapidly
HLSTO	hailstones	PRESRR	pressure rising rapidly
HLYR	haze layer aloft	PRIND	present indications are
		PRST	persist
I			
ICG	icing	**Q**	
ICGIC	icing in clouds	QSTNRY	quasistationary
ICGICIP	icing in clouds and in precipitation	QUAD	quadrant
ICGIP	icing in precipitation		
INTMT	intermittent	**R**	
INVRN	inversion		
IPV	improve	RGD	ragged
ISOLD	isolated	RTD	routine delayed weather
K		**S**	
KDEP	smoke layer estimated—feet deep		
KLYR	smoke layer aloft	SC	stratocumulus
KOCTY	smoke over city	SKC, CLR	sky clear
		SNOINCR	snow depth increase in past hour
L		SNRS, SR	sunrise
LLWS	low-level wind shear	SNST, SS	sunset
LTG, LTNG	lightning	SNWFL	snowfall
LTGCC	lightning cloud-to cloud	SQAL	squall
LTCCCCG	lightning cloud-to-cloud, cloud-to-ground	SQLN	squall line
		SR, SNRS	sunrise
LTGCW	lightning cloud-to-water	SS, SNST	sunset
LTGIC	lightning in cloud	ST	stratus
LTNG, LTG	lightning	STFRA	stratus fractus
		STFRM	stratiform
M		STM	storm
MEGG	merging		
MLTLVL	melting level	**T**	
MNLD	mainland		
MOGR	moderate or greater	TCU	towering cumulus
MRGL	marginal	TOVC	top of overcast
MSTR	moisture	TROP	tropopause
		TWRG	towering
N			
NCWX	no change in weather	**U**	
NPRS	non persistent	UDDF	up and down drafts
NRW	narrow	UPDFTS	updrafts
NS	nimbostratus	UPSLP	upslope

V

VLNT	violent
VR	veer

W

WDSPRD	widespread
WFP	warm frontal passage
WK	weak
WRMFNT	warm front
WSHFT	wind shift
WV	wave

ACRONYMS

AC	—Convective Outlook Bulletin; identifies a forecast of probable convective storms.
AIRMET	—Airman's Meteorological Information; an inflight advisory forecast of conditions possible hazardous to light aircraft or inexperienced pilots.
ARTCC	—Air Route Traffic Control Center, FAA.
CWSU	—Center Weather Service Unit, NWS and FAA.
EFAS	—Enroute Flight Advisory Service (Flight Watch), FAA.
FA	—Area Forecast; identifies a forecast of general aviation weather over a relatively large area.
FD	—Winds and Temperatures Aloft Forecast; a forecast identifier.
FSS	—Flight Service Station, FAA.
FT	—Terminal Forecast; identifies a forecast in the U.S. forecast code.
GOES	—Geostationary Operational Environmental Satellite.
HIWAS	—Hazardous In-flight Weather Advisory Service, FAA
ICAO	—International Civil Aviation Organization.
IFSS	—International Flight Service Station, FAA.
LAWRS	—Limited Aviation Weather Reporting Station; usually a control tower; reports fewer weather elements than a complete SA.
NAWAU	—National Aviation Weather Advisory Unit, NWS.
NESDIS	—National Environmental Satellite Data and Information Service.
NHC	—National Hurricane Center, NWS.
NMC	—National Meteorological Center, NWS.
NOAA	—National Oceanic and Atmospheric Administration, Department of Commerce.
NOTAM	—Notice to Airmen.
NSSFC	—National Severe Storms Forecast Center, NWS.
NWS	—National Weather Service, National Oceanic and Atmospheric Administration, Department of Commerce.
PATWAS	—Pilot's Automatic Telephone Weather Answering Service; a self-briefing service.
PIREP	—Pilot Weather Report.
RAREP	—Radar Weather Report.
SA	—Surface Aviation Weather Report; a message identifier.
SAWRS	—Supplemental Aviation Weather Reporting Station; usually an airline office at a terminal not having NWS or FAA facilities.
SFSS	—Satellite Field Service Station.
SIGMET	—Significant Meteorological Information; an inflight advisory forecast of weather hazardous to all aircraft.
TAF	—Terminal Aviation Forecast; identifies a terminal forecast in the ICAO code.
TWEB	—Transcribed Weather Broadcast; a self-briefing radio broadcast service.
UA	—Pilot Report (PIREP); a message identifier.
WA	—AIRMET valid for a specified period, a message identifier.
WS	—SIGMET valid for a specified period, a message identifier.
WSFO	—Weather Service Forecast Office, NWS.
WSO	—Weather Service Office, NWS.
WST	—Convective SIGMET, a message identifier.
WW	—Severe Weather Watch; identifies a forecast of probable severe thunderstorms or tornadoes.

TABLE 14-3. Scheduled issuance and valid times of forecast products

FORECAST PRODUCTS	TIME ZONE	AREA	ISSUANCE TIME	VALID PERIOD
Terminal Forecast (FT)	Pacific Mountain	—	0940Z	10-10Z
			1540Z	16-16Z
			2240Z	23-23Z
	Central Eastern	—	0940Z	10-10Z
			1440Z	15-15Z
			2140Z	22-22Z
		Anchorage Fairbanks	0440Z	05-05Z
			1140Z	12-12Z
			1640Z	17-17Z
			2140Z	22-22Z
		Juneau	0340Z	04-04Z
			1040Z	11-11Z
			1440Z	15-15Z
			2040Z	21-21Z
		Honolulu	0540Z	06-06Z
			1140Z	12-12Z
			1740Z	18-18Z
			2340Z	00-00Z
ICAO Terminal		All	2340Z	00-00Z
			0540Z	06-06Z
			1140Z	12-12Z
			1740Z	18-18Z
Area Forecast (FA)		Boston Miami	0840Z	09-03Z
			1740Z	18-12Z
			2340Z	00-18Z

TABLE 14-3. (Cont.)

		Chicago Dallas-Fort Worth	1040Z	11-05Z
			1840Z	19-13Z
			0040Z	01-19Z
		San Francisco Salt Lake City	1140Z	12-06Z
			1940Z	20-14Z
			0140Z	02-20Z
		Anchorage Fairbanks	0640Z	07-01Z
			1540Z	16-10Z
			2340Z	00-18Z
		Juneau	0640Z	07-01Z
			1340Z	14-08Z
			2240Z	23-17Z
		Honolulu	0340Z	04-22Z
			0940Z	10-04Z
			1540Z	16-10Z
			2140Z	22-16Z
Transcribed Weather Broadcast (TWEB)	Pacific Mountain		1140Z	12-00Z
			1840Z	19-07Z
			2340Z	00-18Z
	Central Eastern		1040Z	11-23Z
			1740Z	18-06Z
			2240Z	23-17Z
		Alaska Hawaii	None	
Inflight		All	Not Scheduled	See Section 4